PRACTICAL PLACER MINING

Editors
Louis W. Cope
International Placer Consultant
Denver, Colorado

and

Lee R. Rice
Data Technology Services, Inc.
Littleton, Colorado

Published by
Society for Mining, Metallurgy, and Exploration, Inc.
Littleton, Colorado • 1992

Acknowledgements

The editors of this volume are grateful to the several authors who generously took the time to share their knowledge and in so doing, made this book possible.

The Society for Mining, Metallurgy, and Exploration, Inc. is acknowledged with thanks and appreciation for their collaboration and patience in publishing this book.

Printed in the United States of America
by Cushing-Malloy, Inc., Ann Arbor, Michigan

Library of Congress Catalog Card Number 91-66951
ISBN 0-87335-105-3

Preface

This book, *Practical Placer Mining*, is based on a session of the same name at the February, 1991, Annual Meeting of the Society for Mining, Metallurgy, and Exploration. Some additional information has been added, an edited version of the discussion from the floor at the session has been included, and a selected bibliography of placer mining has been incorporated.

The intent of the book is to present a "how to" as well as a "how not to" guide to various facits of placer work. The word "practical" in the title is emphasized in this work.

In several instances, authors of different chapters make the same points using different words. These duplications have not been removed during editing in an effort to preserve each author's work and because learning is based on repetition.

We hope this book will help inform operators and developers and will serve to lower the failure rate of new placer operations.

As readers apply the information presented in this volume, we wish you good luck, buen éxito, bon chance, or whatever language is applicable where you are working.

December, 1991
Denver, Colorado

Louis W. Cope
Lee R. Rice
Editors

Contributors

Editors

Louis W. Cope
International Consultant
Denver, Colorado

Lee R. Rice
President
Data Technology Services, Inc.
Littleton, Colorado

Authors

Louis W. Cope
International Consultant
Denver, Colorado

Charles A. McLean
Mining Engineer
Placer / Alluvial International Consulting
Penn Valley, California

James L. Noble
President
International Resource Development, Inc.
Carson City, Nevada

Douglas J. Piper
Vice President
Ellicott Machine Corporation
Baltimore, Maryland

Mortimer J. Richardson
President
Consolidated Placer Dredging Company
Irvine, California

Joseph R. Wojcik
Branch Manager
Watts, Griffis and McOuat Limited
Denver, Colorado

Table of Contents

Chapter 1

THE HISTORY OF GOLD DREDGING

by Mortimer J. Richardson
President
CONSOLIDATED PLACER DREDGING CO. (CPD)
Irvine, California

EVOLUTION OF DREDGES

Dredging can be traced to early times where it may have been derived by the Phoenicians prior to the 4th century B.C. The first tangible design drawings of a bucket ladder (BL) dredge can be found in the collections of Leonardo da Vinci (1452-1519), which he produced when he was attempting to create better methods of channel and port dredging. The perfection of the system did not commence until after the invention of the steam engine in the early 1800's; providing the impetus to progress from man-powered mud mills, horse drawn along polders or dikes in the Netherlands, England, France and Germany, to steam-driven.

In the 1860's following the various gold rushes in the U.S., Australia, New Zealand, Siberia and South America, with placers in particular, efforts were begun to apply mechanization to achieve deeper digging. There is a recording of a dredge being installed to mine gold in Montana in the 1860's but did not include a processing system. Then in the 1880's in New Zealand, some practical engineers developed a system that is said to be the first bucket ladder mining (BL/M) dredge for mining gold placers.

In studies of the New Zealand gold dredging experience, it is estimated that some 500 dredges were built to mine gold between the 1880's and 1942. A careful analysis of the data, however, shows that very few were profitable. The technology appeared to stagnate in the early part of the century. Two of the BL/M dredge design engineers left New

Zealand in the 1890's to immigrate; one to the U.S. and the other to England. They initiated the design development of what are the two basic categories of BL/M dredges today: The California-type and European-type, respectively. In point of fact, introducing mineral jigs into gold dredges in New Zealand did not occur until the late 1930's when the California Yuba type dredges were introduced.

Most of the technological progress of the BL/M dredge for gold evolved within the U.S. between 1893-1942. Such companies as Yuba Consolidated, Bucyrus Erie, Bethlehem Steel, Marion Steam Shovel, and others to a lesser extent participated in this technology. However, soon the major builder was Yuba, as others dropped out.

Design Influences

There are three principal climatic conditions where gold placers either have been or are now being mined, which influence the configuration of the mining dredge. Other physical characteristics of the environment will create the need for modifications borrowing from one of the other zones.

Permafrost	Alaska, Canada, Siberia, Northern China.
Tropical	South and Central America, Southeast Asia, Africa.
Moderate	Central USA, Southern China, Australia, New Zealand

Permafrost: Frozen ground must first be thawed before dredging and the construction of the dredge must be extra heavy to withstand the stresses. Also, since this usually involves a 6-7 month operating season, maintenance and repair must be limited as much as possible to the winter months, placing an emphasis on durable designs from structure to components. Techniques of thawing the ground have been well established using water injection at ambient temperatures. Other measures including ripping and solar thawing, blasting and steam injection have and are being applied. The California-type, BL/M dredge has proven itself effective in these environments in terms of high production, durability, recovery and low cost of operation.

Tropical Climate: Tropical conditions utilize basically the same dredge systems with enviable results over long periods of time. Dredge construction need not be as heavy but remote conditions dictate the same high reliability to minimize down-time; waiting for replacement parts and repairs. Conditions that are often present include tough clay, buried logs and vines, laterite or other abrasive soil conditions. While they will hopefully not all occur in the same place, they will each require design features to deal with their associated digging and processing problems.

Moderate Climate: Moderate climates have been the scene of large scale gold dredge mining in years past. In the case of the Western U.S., however, little chance exists for any new placers for a variety of reasons not the least of which is the depletion of reserves and intense environmental activism. All of the moderate regions seem to have similar characteristics of soil conditions; *i.e.*, large boulders, river gravels, clay and hard material such as laterite. This has necessitated heavy structural designs throughout the dredge to hold up under severe loadings on a continuous, 12 month/year operation.

Structural Design

Improvements in structural analysis have resulted in sizing reductions in steel construction throughout industry. This has been applied to hulls, superstructure, housing, gearing, motors and about every moving part; particularly with offshore oil platforms and mobile rigs. Unfortunately, some companies have used this more in cost cutting with less emphasis on structural or wear analysis. The result has been seen where dredges are wearing out too early, causing breakdowns and inefficiencies in areas where logistics are poor. Careful scrutiny needs to be given to this aspect, questioning plate and beam thickness and some proof of strength analyses to demonstrate the equipments ability to withstand a 24 hr/day operation. The normal safety factor is not usually acceptable for BL/M dredges, whose structural and component strength evolved through operational experience. "When it breaks, make it thicker," thus the coining of the term "thickening engineer"; dating back to the early part of this century. The proof is in the performance of California-type BL/M dredges that are operating today, overhauled many times but some originally built as early as 1905.

Diminishing Returns on Dredge Size

As the need to dredge deeper became apparent, digging depths and bucket size increased with technological improvements. Mineral Jigs were installed as early as 1914 but it took several years of experimentation before they began replacing riffles entirely, in the mid 1930's. When the size of buckets on BL/M dredges reached 18 ft^3, it was later commented on by a leading dredge engineer, Charles Romanowitz of Yuba Consolidated, that there appeared to be a falling off of production in proportion to the increase in bucket size. He attributed this to the dredge size having reached a point of diminishing returns, where time to raise and lower the ladder, change the heavier buckets and components, all took longer.

This aspect of diminishing returns in sizes of BL/M dredges, has interested me for some time and I have attempted to document the trend in Table 1. The sizes of buckets from 2½ to 30 ft^3, are based on known dredges over a period of years and their production experience. At one time it was thought that the 9 ft^3 dredge was the optimum and bell curves were drawn to illustrate that it reached the top of the curve before diminishing returns began in the larger sizes.

Table 1. Bucket Ladder Mining Dredge Optimum Bucket Size, Cost, and Production Relationships*

Bucket	Depth		Production (x 1000)		Cost†	Factor‡
Size (ft^3)	m	ft	m^3/mo	yd^3/mo	(US$, millions)	f
2½	10.7	35	76.3	100	1.5	15.0
6	15.3	50	160.3	210	3.5	16.7
9	21.4	70	213.8	280	4.5	16.1
14	30.5	100	420.0	550	6.5	11.8
18	38.1	125	328.3	430	15.0	34.9
24	45.8	150	358.8	470	25.0	53.2
30	50.0	164	274.8	360	40.0	111.1

* Examples of size and production are based upon known BL/M dredges.
† Capital costs are either updated estimates or published actual costs.
‡ Factor f = cost in $/$yd^3$ of monthly throughput, showing a relationship of cost effectiveness; *i.e.*, the smaller the number the more efficient the operation.

The experience of Pato Consolidated Gold Dredging on the Nechi River, Antioquia, Colombia has provided substantial proof of optimization using 14 ft^3 dredges. Several factors made their high production and low cost operation possible; *i.e.*, standardization with five dredges of the same size operating from a single support facility. Nevertheless, it has been demonstrated that optimization does occur at that size. Total production by PATOs dredges on the Nechi River, from 1932-74, has been quoted in excess of 763 million m^3 (1.0 billion yd^3).

In Table 1, further data is furnished on each size dredge that helps to quantify the experience of production as it relates to the capital expenditure. Missing is the cost of operation and its tendency to reduce as production increases. However, examining those costs has not produced a trend that is useful and if anything, has shown a degradation (of cost/m^3) after the optimum size of 14 ft^3.

REFERENCES

Bogdanov, Y. I., 1990, "Placer Mining Technologies in the U.S.S.R.," Alaska Miners Association Minerals Symposium, Anchorage, AK

Burton, A. K., 1990, "The Project Financing of Small to Medium Mines: A Banker's Perspective;" *Financing of Gold Mines Seminar*, Southern California Mining Section of SME, Irvine, CA.

Cleaveland, N., 1982, "Circular Jig," U.S. Patent No. 4,310,413.

Cleaveland, N., 1973, "BANG! BANG! IN AMPANG, Dredging Tin During Malayas Communist Emergency," Symcon Publishing, Irvine, CA.

Kvint, V., 1990, "Go East, Young Man," *Forbes Magazine*, New York, Nov., p. 234.

Anon., 1990, "The Klondike -- Soviet Style," *Engineering and Mining Journal*, Nov.

Richardson, M. J., 1988, "Finding & Developing Placer Gold Mines," *International Congress on Dredging Mining Systems*, Ellicott Machine Corp., Baltimore, MD.

Richardson, M. J., 1987, "Site Evaluation Criteria, Exploration and Development of Small-Scale Placer Gold Mines," *Small Mines Development in Precious Metals Conference*, SME, Reno, NV, p. 69.

Richardson, M. J., 1986, "Evaluation/Decision Process for Small-Scale Placer Gold Mining;" *Mining Magazine*, London, April, p. 312

Richardson, M. J., 1985, "Placer Gold Mining; A Return to Basics," *World Dredging Magazine*, July, p. 12.

Richardson, M. J., 1984, *Handbook of Alluvial Gold Evaluation Methods*, Consolidated Placer Dredging, Irvine, California, 300 pp.

Richardson, M. J., 1983, *Handbook of Mineral Jigs*, Consolidated Placer Dredging, Irvine, CA, 200 pp.

Richardson, M. J., 1982, "South American Dredge Mining," World Dredging Magazine, Irvine, CA, July, p. 12.

Richardson, M. J., 1980, "Dredge Mining Placer Gold in the 1980s," *WODCON IX Proceedings*, Western Dredging Association, Arlington, VA, p. 809

Chapter 2

PLACER SAMPLING

by Joseph R. Wojcik
Branch Manager
WATTS, GRIFFIS AND McOUAT LIMITED
Denver, Colorado

INTRODUCTION

A great deal more is involved in a sampling effort directed at a placer deposit than physically capturing a quantity of representative material and extracting the desired component. The distribution and dimensions of the sample sites must be chosen after consideration has been given to the type of deposit and the concentrating mechanism. Interpretation of the sampling results is likewise an extremely important part of the sampling program.

SAMPLE SPACING

The interpretation begins with the selection and design of the sampling program. A decision regarding genesis of the deposits and their geometry must be made in order to select a suitable technique and sampling pattern. Deposits differing genetically will have differing distributions and degrees of concentration.

Table 1. Placer Deposit Characteristics

Genetic Type		Horizontal Distribution	Vertical Distribution
Residual	(Eluvial)	Blanket, Isotropic	Single Layered
Residual	(Fluvial)	Sinuous, Anisotropic	Stacked
	(Aeolian)	Parallel, Sinuous, Anisotropic	Single Layered
Weathered	(Enriched)	Distribution Overprint	Complex
Glacial	(Fluvial)	Sinuous, Anisotropic	Complex
	(Marine)	Blanket & Sinuous Parallel	Complex
	(Aeolian)	Blanket, Parallel Sinuous	Single Layered

Since fluvial (water laid) placer deposits are anisotropic (highly elongated) with width to length ratios exceeding 1:10 in some cases, the drill hole spacing should reflect that anisotropy. I like to use a ratio of 1:5. With such a ratio, holes along drill lines 224 meters apart would be spaced 48 meters apart. Optimum spacing across the deposit can be estimated from a semi-variogram.

For prospecting, the drill hole spacing can be estimated from the expected width of any pay horizons and the expected anisotropy. If a pay horizon of 200-meter width is expected with a deposit anisotropy of 1:5 then the drill hole spacing would be 100 meters apart along lines 500 meters apart.

SAMPLING METHODS

Other than churn drilling, the techniques for sampling placer deposits are only limited by one's imagination. Some techniques used today are shown in Figures 1 and 2, and include:

- Banka Drill
- Becker Hammer Drill Reverse circulation
- Rotary Reverse Circulation
- Yost Klam Drill with Caisson
- Bucket Auger
- Hollow Stem Auger
- Vibracore
- Hawker-Siddely Resonance Core
- Backhoe

- Ditch Witch
- Shaft and Caisson

Fig. 1. A hand-operated Ward Drill at a remote jungle site in Costa Rica.

The primary purpose of a drilling and sampling program is to estimate as closely as possible the value of the sampled ground from the sample. Much has been said and written about the minimum size of sample that is necessary to estimate the value of the ground with a high degree of confidence. Large samples over a large tract are impractical and excellent estimates can be produced from a carefully operated drilling program.

Fig. 2. A modern, sophisticated, reverse-circulation drill at a site in Nevada.

INTERPRETATION OF GOLD EXPLORATION DRILLING PROGRAMS

Gold Weight

In the precious metal/alluvial industry, the accepted drilling tool has been the Keystone (churn, cable) drill and certain practices have been used to calculate an average value for the ground from the Keystone sample. Since value of placer ground is expressed as contained gold weight per unit volume, these parameters must be determined for the small sample and extrapolated to the unit volume, usually cubic meters or cubic yards. Ideally, a casing is driven into the ground, the material displaced into the casing retrieved and its gold content determined. The weight of gold recovered can be weighed and, assuming that the recovery technique is efficient, the gold weight may be considered accurate with a high degree of confidence.

Estimating techniques using counts of gold particles in different standard size ranges lessens the degree of confidence in an individual sample interval weight but since the gold from the hole is weighed in total and the total hole was panned by one panner, the adjusted weights should be consistent within a range. Correcting the weights estimated for each sample interval by the ratio of actual weight per total estimated weight in each hole gives a very close approximation of the true interval weight; *i.e.*, the weight of gold in a **full** cubic yard of material is projected from the weight of gold in the sampled fraction of a cubic yard. Various schemes to estimate weight by color count use 3,4 or 5 size ranges and the sizes 1,2 and 3 are not of equal average weight in all of the schemes. Any of the schemes are acceptable so long as the panner is consistent. Never use more than one panner on one hole. A panner's estimates should sum to 60-80 percent of the actual weight and will usually vary only 5 percent for an individual experienced panner. Wide variations are recorded when large colors are estimated and large colors should be weighed individually.

Sample Volume

Assuming an acceptable estimate of the gold weight has been recorded, the volume from which that gold was separated must also be estimated. A bulk sample volume can be calculated from measurements of the dimensions of an excavation but a drill hole is not so easily measured. Two techniques are used to estimate the in-place volume of a sample. One technique requires that the sample be deposited into a calibrated container at the surface and the measured volume recorded. This measurement is of a disturbed sample and is not in-place volume.

Coarse, granular soils can swell such that their loose volumes are more than 30 percent greater than their in-place volumes. Allowance must be made for swell when using loose measurements in the value calculations. The other technique involves measuring the amount of sample displaced into the casing after the casing has been driven ahead incrementally. This measurement more closely approximates the in-place volume from which the gold is eventually separated

The casing used most commonly on Keystone drills around the world has been of nominal 23.6 mm (6 inch) diameter but is in fact 22.6 mm ($5^3/_4$ inches) in inside diameter and is driven with a 29.5 mm (7½ inch) shoe tapered 45° toward the center. The taper is to displace all of the material cut by the outside edge of the shoe into the casing. If this were really achieved, a unit advance of the shoe would result in 1.7x the unit distance of sample inside the casing (10 cm advance would produce 17 cm sample inside). This theoretical relationship is rarely experienced and various explanations have been offered for the discrepancy between actual "core rise" and theoretical "core rise." The discrepancy only becomes a problem when calculations of values are related to theoretical volumes. The following discussion must be presented in English units for clarity.

Early in the twentieth century, Colonel Radford settled on a theoretical volume for a 7½ inch diameter drive shoe such that 100 feet of hole contained one cubic yard of material. The equivalent diameter of a 100-foot long core of one cubic yard volume is 7 inches. These two measurements have been used in sample estimations as the Radford or Keystone factor. Radford explained the smaller core diameter as resulting from rounding of the edge of the drive shoe. The Radford factor of 7 inch theoretical core diameter closely approximates the core rise experienced in the field which commonly ranges from 1.4 to 1.5 times the unit advance. I suggest that, given the way the value of the ground was calculated by Radford and many following engineers, the 100 ft/yd^3 and 7 inch theoretical diameter were chosen largely for convenience and only relate fortuitously to the mechanics of the casing shoe action in the alluvial soils.

Calculations

The American technique of drill log calculations converts each interval gold weight to an equivalent for a theoretical volume for the increment of advance. In this technique, it is assumed that the gold recovered from an individual sample is 100 percent of the gold originally present in that sample and the weight can be adjusted upward or downward in an amount that is proportional to the volumes. If a smaller than expected core rise is measured, the weight of the gold is adjusted upward for the interval by the ratio of theoretical core rise to actual core rise as shown in equation 1:

$$Corrected\ Weight\ of\ Au\ (mg) = \frac{Weight\ of\ Au\ (mg) \times Drive\ Factor^{*}}{Rise^{**}} \quad [1]$$

* Expected core rise for diameter of drive shoe used.
** Actual core rise.

Likewise, if a greater than expected core rise is measured, the weight of gold is adjusted downward by the same ratio. All gold weights in a drill hole are then proportioned to the equivalent weight for a theoretical core diameter which in the case of the Radford factor is 7 inches. The overall value in weight per volume of the sample interval, for the entire hole, is calculated from Equation 2:

$$Weight\ of\ Au\ (mg/yd^3) = \frac{Weight\ of\ Au\ (mg) \times Constant^{*}}{Depth\ (ft.)^{**}} \quad [2]$$

* Ft/yd^3 for diameter of drive shoe used.
** Depth in feet drilled to produce the sample.

If a theoretical core diameter of 7½ inches is used, the ratio is 88 feet per cubic yard/interval, feet. When using the 7½ inch theoretical core diameter, the individual sample gold weights will be adjusted to a theoretical rise of 1.7x instead of the 1.48x used with the Radford factor. So long as individual sample gold weights are adjusted upwards or downwards to a theoretical volume and the equivalent length of hole per cubic yard is used to calculate the value of the sample or hole, any theoretical diameter can be used. The 100 feet per cubic yard is a convenient multiplier and the 1.48x core rise factor results in fewer upwards adjustments of gold weights than the 1.7x core rise factor. With a $5^3/_4$ inch theoretical diameter, the theoretical core rise factor would be 1.0 and virtually all sample gold weights would be adjusted downwards but the multiplier would be 150 feet per cubic yard and the calculated value of gold per cubic yard would be the same.

KEYSTONE SAMPLE CALCULATIONS FROM CORE RISE

The apparent core diameter of 6.8 inches to 7.0 inches is not a function of rounding of the shoe edge but rather a result of the interaction of the coarse granular, non-cohesive soils and the geometry of the casing drive shoe. We have long recognized that, when driving casing in tight, clayey or cemented ground, core rises frequently were measured at 1.7x advance. This performance is only experienced in cohesive soils. Sandy soils give core rises between 1.5x and 1.6x. The casing drive shoes have an inward taper of 45° but the internal angle of friction of the non-cohesive soils is a function of the mean particle size and degree of sorting. In very coarse, loose gravel, the internal angle of friction approaches 40°, the shear force transmitted by the 45° drive shoe face is nearly vertical and the core diameter will be that of the inside of the drive shoe. Cohesive soils will be sheared by the drive shoe edge and a full diameter core will be recovered.

An alternate method of calculating the value of the ground from drill samples simply divides the weight of gold recovered by the volume of the sample. When the volume used is the volume calculated as the product of the core rise and the casing internal cross-sectional area, the resulting calculation should be the most accurate estimate of all methods. When the volume used is a measured loose volume, an appropriate adjustment must be made for the swell. An old argument that the swell in the loose volume compensated for fines that were lost as slimes in the wash water can not be supported in fact. It would require a slimes content of up to 35 percent to compensate totally for swell in some very coarse material, the very material in which slimes content would be expected to be the lowest.

In order to estimate value using a weight/volume calculation, two factors must be empirically determined:

1. Swell factor
2. Theoretical core diameter.

The swell factor can be estimated by excavating a measured volume into a calibrated container and the theoretical core diameter can be estimated from an analysis of the distribution of sample volumes adjusted for swell. The distribution should be normal and the theoretical core diameter one that will produce the sample volume at the mode of the distribution curve. This technique is applicable to the Keystone as well as Becker drilled holes and to caisson holes, shafts and other bulk sampled sites. It is imperative, however, that all of the sample displaced into the casing be delivered to the surface collection system. Inefficient and inexperienced drilling cannot produce a sample from which a reliable estimate of value may be calculated. The warning that, "no amount of mathematical manipulation can make a good sample out of a bad one" is never more true than in the evaluation of alluvial deposits.

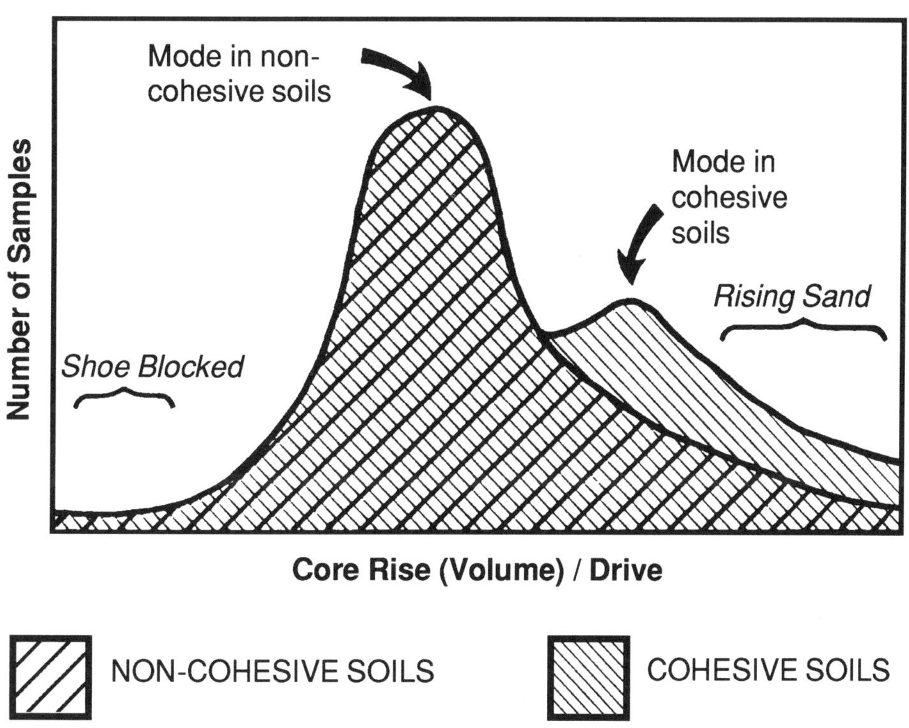

Fig. 3. Distribution curves of core rise.

RESERVES

Interpretations of results from sampling programs in alluvial deposits are often done without thoroughly understanding the deposits and the reports are erroneous or ambiguous at best. At times this is a deliberate attempt to defraud the potential investors, but most often it is simply inexperience.

Alluvial deposits are seldom single deposits but rather the aggregate of superposed and/or contiguous discrete smaller deposits with similar though different genesis and distribution.

Alluvials should always be studied in cross section before correlations are entered on the plan view. In the cross sections, correlations can be based on contacts, clay content, grain size, sorting, gold content, nature of contained gold, accessory mineral suite or any other distinguishing characteristic or combination of characteristics. If necessary, the plan view can be constructed initially on several planes to depict the trends of mineral-bearing ground. Both geologic and gold-content-per-unit-volume boundaries will be found to be appropriate at various locations. Ultimately the inventory of valuable mineral may be compiled in a manner consistent with the mining method.

Reserves may be calculated on cross sections projected along trend or in plan using polygons, triangles, rectangles or moving averages so long as all holes within the correlation are included. The resulting estimate will usually be conservative because sample recoveries less than perfect will more often underestimate the true value than they will over estimate. That is the loss of one grain of the valuable mineral will affect the apparent grade of the sample far more than the loss of one grain of the matrix. For this reason, cutting (reduction) of high values should be done prudently, if at all. Cutting high values to the mean plus two times the standard deviation is a commonly used technique. Sometimes it is not appropriate. A variation of the mean plus two times the standard deviation is to cut the highest 2.5 percent of the sample values to the 97.5 percentile <u>and</u> raise the lowest 2.5 percent of the sample values to the 2.5 percentile, placing all values within the 95 percent range. Another useful cutting technique is to cut the highest samples to the projection of the cumulative frequency distribution curve within the grade range classification.

The method used at the Valdez Creek gold mine in Alaska was to limit the maximum number of ounces estimated from an individual hole or a particular section or to limit the area of influence of a high grade sample.

Most importantly, in an estimation of the mineral inventory, always strive to make the best estimate of the mineral content in situ. Future engineers and geologists can apply whatever factors and recovery, extraction or dilution that are appropriate for the mining plan under consideration. Alluvial deposits in particular pass through many drilling campaigns and often many owners before they can be developed successfully. Careful calculation and reporting of sampling results and procedures will prevent errors from occurring in subsequent evaluations.

SUMMARY

Proper evaluation of alluvial deposits is accomplished by,

1 Careful selection of a:
- Genetic Model
- Sampling Pattern
- Sampling Technique

and by,

2 Intelligent interpretation of the results.

Care in sample collection, measurement and recording will be rewarded by accurate extrapolations. The motto should always be "no amount of mathematical manipulation can make a good sample out of a bad one."

REFERENCES

Bronston, M.A., 1990, "Offshore placer drilling technology: A case study from Nome, Alaska," *Mining Engineering*, Vol. 42, No. 1.

Tinsley, C.R., 1990, "Why Alluvial Gold is Unbankable," *Project Finance*, Indosuez, Ltd., Sydney, N.S.W.

McCallum, R.W., 1967, "The Ward Hand Drill," *Mining Magazine*, Vol. 116, No. 3.

Woolsey, J.R. and Noakes, B.G., 1990, "Remote placer drill and application for reconnaissance of marine precious metal placers," *Mining Engineering*, Vol. 42, No. 9.

Stone, J.G., Mejia, V.M. and Newell, G.T., 1988, "Using diamond drilling to evaluate a placer deposit: A case study," *Mining Engineering*, Vol. 40, No. 9.

Cleveland, G. and Bressler, J., 1990, "Drilling Methods Used in Exploration of the Valdez Creek Paleochannel Deposit, Alaska," Dredging and Placer Mining Conference VII, Reno, Nevada, SME, Feb.

Chapter 3

SAMPLES, BULK *vs.* DRILL

by Louis W. Cope
International Placer Consultant
Denver, Colorado

INTRODUCTION

Various methods for sampling placer deposits have been mentioned by Wojcik in Chapter 2. Here, a brief description and comparison of these methods will be given as related to bulk versus drill sampling. The advantages and disadvantages inherent to each will also be given.

The means of sampling can be grouped into bulk or drill samples. As a generality, drill samples over a five-foot length will run from $^1/_{20}$ to $^1/_{50}$ cubic yard, depending on the diameter of the drive shoe on the drill casing. Bulk samples can be expected to be in the range from one to many cubic yards each.

SAMPLE METHODS

Bulk

Bulldozer-dug trenches and dozer or front-end loader pits are two methods to excavate large bulk samples. Smaller bulk samples can be obtained by digging small dimension shafts through the deposit, by hand or caisson-like drills. It is not recommended to split samples by alternate bucket-loads or by other means, as the effect of a nugget falling into one or the other fraction will seriously alter the sample.

Drill

As was mentioned in Chapter II, drills used in placer sampling include the Banka hand-operated, churn, auger, reverse circulation and vibratory drilling systems. For sample integrity, the holes should have a casing, or the drill stem should be nearly the hole diameter, thereby serving as a casing. Pits or trenches can be excavated, then a channel sample taken from the side of the hole to be tested for gold content. These channel samples can be of a dimension that are equivalent in size to a drill sample.

SAMPLE REDUCTION

Bulk Samples

As the title implies, a bulk sample is one of considerable weight and volume. For this reason, large-size process machinery is required to reduce a bulk sample to a gold-black sand concentrate. For excessively large samples, pilot-scale equipment is necessary, and this could be equivalent to a small-size operating plant. Sluices can be used, but this method normally loses the fine-size gold particles.

Drill Samples

The small volume and weight of drill samples allows them to be reduced to the heavy fraction by hand panning, small sluices, or special motorized small-throughput machines. Model names of these machines include the Gold Miser, Gold Saver, Prospector, and Bowl Separator. Normally all rough concentrates produced by the machines named above have to be finished by hand panning.

GOLD DETERMINATION

To determine the weight of gold in the concentrate, the larger gold particles which physically can be removed, are removed. The remaining finer gold is then recovered as described below, and is weighed on a scale with a minimum of one milligram sensitivity.

In no case should fire or atomic adsorption assays be made on the whole concentrate, as this shows the total amount of gold present, not just the amount recoverable to gravity processing methods. However, after the recoverable gold is removed, an assay of the black sand residue will show if grinding with amalgamation is necessary and feasible.

Amalgamation

To amalgamate the gold in the concentrate, mix it with a small amount of caustic, to clean oils off the gold surfaces, and mercury in a thick water slurry. The mixture is agitated, then the amalgam is removed. The mercury is dissolved by dilute, warm nitric acid, and the gold residue is washed, dried, annealed, cooled, and weighed.

Blowing

The gold can be separated from heavy minerals in a dried sample. It is put on a flat surface and someone carefully blows only hard enough to move the heavy minerals, leaving the gold on the surface. This method is mainly used with small samples.

Mechanized

There are small motor-driven machines on the market which are useful in separating gold out of concentrates. These are inclined, dish-shaped containers with riffles set in a spiral configuration. These units are very useful in working with larger volume concentrates from bulk samples.

INFORMATION

While sampling is being carried out, following are items of information and data which should be recorded.

Ore

The volume of each raw sample is a vital piece of information for comparison with the amount of gold recovered from it. An estimate of the quantity of each size fraction of material in the ore: i.e., boulders, cobble, gravel, sand and silt; should be noted. Knowledge of the difficulty of digging or drilling, or if the muck is cemented, helps plan the mining method and sequence. Laminations of the gravel, ground water quantity and location, distance from surface to bedrock as well as to any false bedrock are all vital bits of information. For plotting samples points, a good topographic map of the sample area is necessary. Government topographic maps are not good enough due to their small scale and contour interval lines being in the 20- to 50-foot range.

Gold

The quantity of gold to compare with the volume of gravel it is recovered from is the purpose of sampling. A screen analysis of the gold particles is very useful to know. The shape and surface condition of the gold particles should be noted. The fineness of the gold can be determined in a laboratory.

Black Sand

Heavy sand in the concentrate should be weighed to determine its percentage of the gravel ore. The percent of the magnetic fraction, which is largely magnetite in most cases, should be determined.

General

The climate, vegetation, and soil cover should be noted. In addition, notes should be taken on infrastructure items such as site access, power, water, labor availability, etc.

ADVANTAGES AND DISADVANTAGES

Bulk Sampling

Advantages: The advantages of bulk sampling include a good view of the gravel in place, and knowledge of the amount of force required to excavate it. Some of the same digging machines used in bulk sampling are used in production. The large samples obtained minimize "nugget effect" and errors

Disadvantages: The fact that pits or trenches may not reach bedrock in all cases is a serious disadvantage. It is often difficult to accurately measure a pit due to curves in the excavation if by a back hoe, and sloughing from the sides. If ground water is encountered, sample integrity is lost. Larger samples require larger transport and processing equipment, and the situation becomes more complicated. As unit costs of each sample are high, the exploration cost to delineate reserves is very expensive.

Drill Sampling

Advantages: Historically, drilling has been the sample method of choice and has been proven by operation in most cases. Drilling gives greater coverage in developing gravel grade and reserves at less unit cost, the profile of bed rock, and indications of enriched and barren areas.

Disadvantages: Drilling gives a small sample which aggravates the nugget effect. At times, boulders will necessitate abandonment of uncompleted drill holes. Vibration from driving casing or from a reverse circulation hammer-drill may cause gold to migrate downward into the gravel. Drill rigs require at least minimal access roads and drill-pads.

RECOMMENDATIONS

At best, placer sampling leaves much to be desired. For this reason, the more samples taken, the better the statistical average the results will be. For sampling, I recommend a combination of drilling and pitting. Drill close-spaced holes in "fences" across the trend of the placer deposit, with these lines of holes at a relative large distance apart, as suggested in the previous chapter. Then, a few pits should be dug to confirm the type, proportions and segregation of material in the deposit.

Chapter 4

SMALL SCALE PLACER MINING

by James L. Noble
President
INTERNATIONAL RESOURCE DEVELOPMENT, INC.
Carson City, Nevada

INTRODUCTION

A small placer operation can be defined as one with a throughput capacity of 250 cubic yards an hour or less. The success of small placer operations after the values have been established, depend entirely on having the proper design. There are various types of placer deposits including marine, beach, desert, stream, bench, underground and even frozen placers. Each is unique and requires special considerations.

Size distribution of alluvial placer ore varies from sand to large boulders. When boulders occur, they can become a problem to all types of excavating to a greater or lesser degree. Clay content and cementing are common in placer deposits. Some clay disperses easily in water while others, if not handled properly, will form clay balls that pick up gold and carry it to tailings.

Overburden, depth of bedrock, vegetation, and ground and surface water must all be considered when designing a plant. Any equipment chosen should be capable of processing large volumes of material at a high efficiency. There are usually five steps to processing placer gravels, as follows:

1. Excavating
2. Feeding the Plant
3. Washing and Classifying
4. Concentrating
5. Tailings Disposal

EXCAVATION

Underwater

Underwater excavation is usually accomplished by the use of a floating dredge. The most common type is a suction unit which uses a pump to lift the gravels. When the material to be excavated is slightly cemented, a cutter head is attached to the intake to assist in digging. In smaller operations, a jetted suction dredge can be used. In this method water is pumped into the suction line creating the lift needed, which eliminates the use of a pump and high wear to it.

A potential problem with underwater excavating is the downward migration of gold as digging underwater stirs up the muck. Also, digging in muddy water increases wear and tear to the machinery.

Dragline

Draglines are useful in excavating both dry gravels and those which lay underwater. Some drag lines have booms up to 100 feet or more giving them a large digging radius. They can feed a process plant which is floating in the pond made by the drag line, usually in excavation of the auriferous gravel.

Backhoe

Backhoe or track excavators have a hydraulically operated arm with a bucket. Because of their limited reach when used on dry land, they must be close to the floating process plant or mounted on the same barge as the plant. When a plant is land-based, it should be portable enough to be moved forward frequently as the backhoe advances.

Bulldozer

Bulldozers are popular with placer miners when it comes to stripping vegetation and removal or overburden. Bulldozers with rippers are used to break up cemented gravels, as well as to move muck as required.

Front-End Loaders

Front end loaders and hydraulically operated front buckets are used to excavate loose or ripped gravel. The rubber tired front end loader is much faster than a track loader, however, it lacks efficiency when digging tight in-place gravels.

Hydraulic Monitors

Hydraulic monitors are used where a considerable amount of water is available. The gravels are excavated by a stream of high pressure water through a nozzle directed at a bank. The muck is loosened by direct impact of jets of water, or is undercut until caving occurs.

The required water pressure is applied to the bank by the monitor. Water pressure is a result of natural head or by diesel or electric pumps. Under the right conditions, hydraulicking can be very efficient in moving large amounts of muck at a low cost. The large amount of water and tailings create serious problems that may be environmentally unacceptable, unless settling ponds of considerable size are built.

FEEDING THE PLANT

Hoppers

Hoppers are used when it becomes necessary to dump gravels directly into the plant from the bucket of an excavator, dragline, backhoe or front-end loader. The hopper must be of sufficient strength to withstand the abuse of direct dumping while having a large enough capacity to maintain an even feed rate to the plant. A grizzly may be provided to reject oversized boulders which can cause damage to the process plant.

Conveyors

Conveyers are very efficient when moving muck from the excavator to the plant. Rubber belting, moving on idlers and driven by a power-driven head pulley can be as much as several hundred feet long when required. Conveyors are relatively inexpensive and can reduce the frequency of moving the plant as the excavation advances.

When properly designed, conveyors are very dependable except when the ore contains boulders, or water content is excessively high allowing the muck to run on the belt.

Pumps

Pumping of slurry is sometimes necessary to feed the plants. The pumps are usually centrifugal-type, either rubber lined for sand or hard iron for gravel. Because of high wear and operating costs, slurry pumps are avoided wherever possible. When they are used, dredge-type pumps are specified. These have open impellers and therefore have low efficiency.

PROCESS PLANTS

Pilot Plants

Pilot plants are very useful to verify exploration information. They should use the same type of machinery as will be used during production. This will enable the operator to anticipate the recovery efficiency of his future production plant. A typical test plant will process about 5 cubic yards per hour and consist of a grizzly, feeder, trommel, jig, and stacker conveyor. The plant should be portable, easy to move, set up and operate. Land based plants can be fixed in place or portable as needed.

Land-Based Plants

Land-based plants can be fixed in place or portable, as needed. The fixed plant is usually installed near a large deep deposit where enough gravel is close by so as not to incur high transportation costs.

The plant is fed by conveyors from land or the material is pumped from a floating dredge digging under water. The portable land-based plant is usually skid mounted and is moved frequently following the excavator as it advances along a narrow or shallow pay streak.

Floating Plant

A floating plant as shown in Figure 4, is mounted on a barge and has similar equipment to the land base plant such as trommel and jigs. It is fed by a drag line or back hoe. The plant will advance forward following the excavator on the dredge-made pond while discarding the tails by a stacking conveyor or chutes behind it, to backfill the excavation.

Clean Up

Clean-up of concentrates from the process plant is either done on site or taken to a secure area away from the mine site as security is always a serious consideration. The concentrates can be upgraded by further gravity concentration, amalgamation, or a combination of these or other methods.

Fig. 4. A 50 yd^3 capacity floating dredge in South America. It is fed by a backhoe mounted on the same barge.

DESIGN PARAMETERS

Mining System

A mining system should be selected to fit local conditions and operate efficiently at low cost. Two items must be taken into consideration when sizing the system.

a. The minimum size is determined by the values versus production cost per cubic yard to produce a profit.

b. The maximum capacity is determined by the amount of reserves and capital cost of production equipment.

Process System

The process system must be designed to accommodate the mining equipment chosen. If possible, flow through the system should be by gravity as much as possible to eliminate pumping. Always maintain a large tailings disposal area. If not, the plant will be covered in a short time by tailings, due to "expansion" of the muck in digging, treating and depositing it.

COSTS

Capital Cost

Capital costs for small plants of 250 cubic yard an hour or less can be estimated for budget purposes at $3,000 per cubic yard per hour. This is for the process equipment only. If a floating plant is required, it can cost as much as double this figure. Excavating machinery is also an additional charge.

Operating Cost

Operating costs are dependant on local conditions. Usually the operating cost of a 100 cubic yard an hour plant will be from $3.00 to $5.00 per cubic yard including depreciation. As can be expected, the larger the plant capacity, the lower the cost per cubic yard of gravel treated.

Chapter 5

BUCKET LADDER DREDGES

by Charles A. McLean
Mining Engineer
PLACER / ALLUVIAL INTERNATIONAL CONSULTING
Penn Valley, California

INTRODUCTION

Bucketline dredges are machines for excavating large relatively unconsolidated or weakly consolidated materials from below water level. They are at the upper end of the capital cost range and the lower end of the operating cost range for placer mining machines and have digging capacities, typically, of from 100 cubic yards of bank measure material per operating hour for a shallow digging dredge carrying 2 cubic foot buckets to 1500 cubic yards per hour for a large deep digging dredge with 33 cubic foot buckets. Bucketline dredges used in the mining industry have their mineral processing plants mounted aboard the digging hull.

Since dredges are high capital cost items, they require large ore reserves, typically at least ten years. Because of their high capital cost they are usually operated around the clock and for as many months of the year as weather conditions will permit. Typically a well run dredge operating year round would be expected to operate for 600 hours per month. Dredges operated in cold climates cannot operate when the weather freezes the dredge pond or the gravels, and dredges operating at sea cannot operate when high waves might batter the digging ladder against the bottom or upset the mineral extraction process.

BUCKETLINE DREDGES

There are two types of bucketline dredges, the spud dredge anchored at the stern by a vertical spud plunged into tailings to hold the dredge against the face as it digs, and the headline dredge pulled up against the digging face by a headline. The spud dredge is able to dig tougher ground due to the rigidity of its anchoring arrangement but requires that coarse tailings be directed to the stern of the dredge to within a few feet of the bottom of the dredge hull, for the spud to be anchored in.

The headline dredge is tied to an anchor well ahead of itself by long heavy line and the elasticity of the line protects the bucket band to some extent to shock overloading from difficult digging. Where based inland the headline dredge must have the ground cleared ahead of itself for an unimpeded swing for the headline, and the sometimes violent movement of the headline across the face as the dredge digs makes it impossible to do stripping or other earthwork on the working face of an operating headline dredge. A sketch of a bucketline dredge is shown in Figure 5.

A bucketline dredge digs by swinging the bow cutting end of the dredge from side to side with side lines. The digging starts at the top the ground, often several feet above water level for inland dredges. At the end of each swing the bucket ladder, which is attached to the dredge and pivoted near the upper tumbler, is lowered several feet to swing again across the face. The process is repeated to the bottom of ore. At its greatest depth for digging, the ladder is depressed to about 45° below horizontal.

Modern bucketline dredges are limited to practical digging depths of something less than 200 feet. The tensile strength of the manganese steel used in the buckets determines the practical limit of digging depth in present day dredges. Deep digging bucketline dredges require idlers beneath the bucket ladder to control the loose catenary loop of buckets below the ladder and reduce tensile stress on the bucket band.

The design of dredge buckets is complex. The base of the bucket is a part of the power train to the digging face, the bowl of the bucket is a haulage container, and the lip is a digging device dependant for its geometry on swing velocity, bucket band velocity, the length of forward step and power available, and the composition and toughness of the material being dug. Obviously there is interplay between the design of the bucket and the design of the other items of machinery used in association with it.

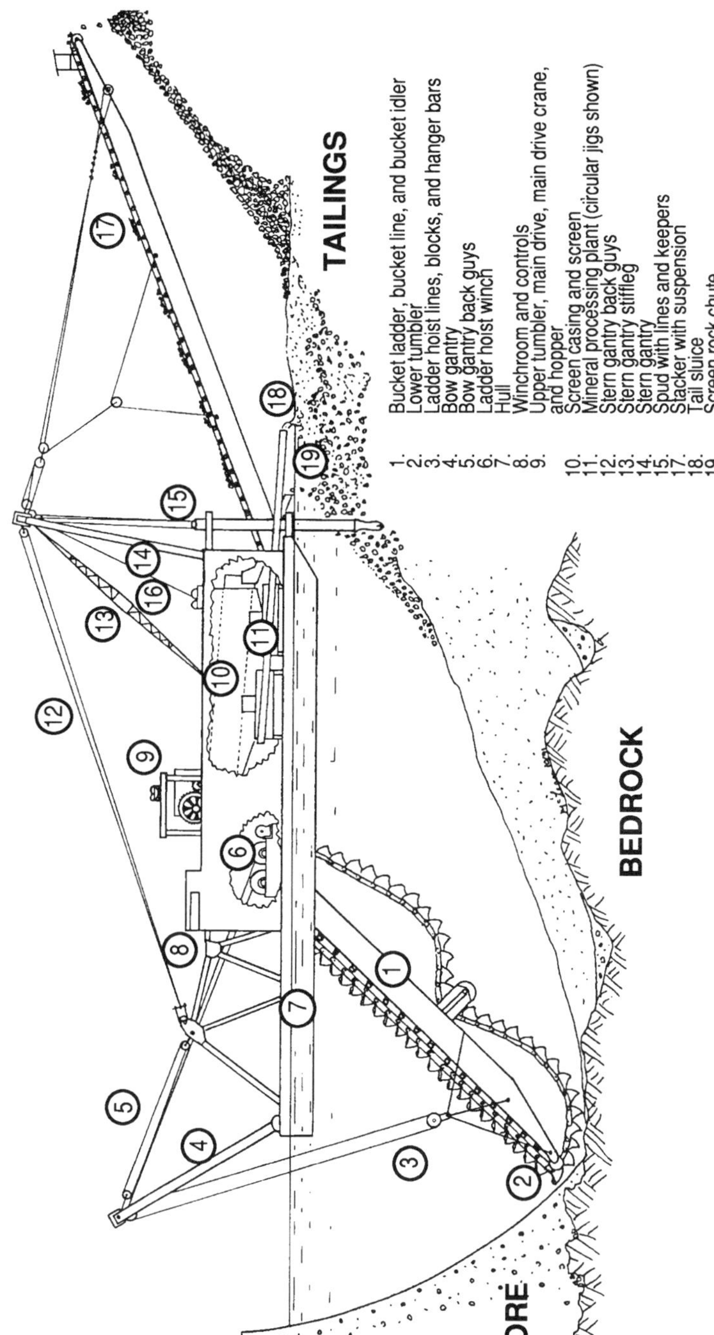

Fig. 5. Components of a California-type bucket ladder dredge.

The dredge is swung during the digging process and moved about the dredge pond by sidelines, the port and starboard bow lines and the port and starboard stern lines. These lines are anchored ashore on inland dredges. With spud dredges the bow line winches swing the dredge from side to side pivoting the dredge about its spud at the stern and the stern lines are only used for maneuvering. Some headline dredges hold the stern in place with tight stern lines and pivot about their sterns operating in this respect in much the same manner that a spud dredge does. Other headline dredges synchronize the stern and bow lines so that the two side lines pull together and the dredge digs back and forth across the face in a straight line rather than an arc. Synchronized lines can dig wider faces in one cut (dependant on winch drum size) decreasing the amount of time spent moving the dredge, but are not as maneuverable.

The dredged material is cut from the face by the side of the bucket as the dredge swings back and forth and is elevated from the lower tumbler in the traveling band of buckets riding on ladder rollers, much as a conveyor belt rides on its idlers. At the upper tumbler the material is usually dumped into an inclined hopper where it passes directly into the dredge's ore treatment plant. Some material, usually somewhat concentrated, remains in the bucket after the initial dump and is caught in the saveall in the ladder wellway. Savealls recover from 3 percent to 10 percent of a dredge's production depending on the design of the bucket and hopper, and the quality of the saveall.

PROCESSING

Typically, separation of valuable heavy minerals aboard bucket ladder dredges starts with the washing and classifying of bank run material dumped from the buckets into a hopper and immediately into a screen although some dredges, requiring greater height for their upper tumblers, will dump onto a grizzly which protects the screen from coarse rock and diverts coarse rock to some other area for subsequent disposal. The screen, either a trommel where the material can be broken up under high pressure water, or a multi-decked shaking screen sizes the feed to the plant. The undersized material almost invariably contains the valuable heavy mineral form concentration and gravity separation methods are used because of their low unit cost.

Older dredges for gold, tin and gems used gravity to distribute material from trommels to riffled "tables" for gravity recovery of coarse valuable mineral. Those older type dredges had to be shut down from time to time to clean up their "tables." Those old bucketline dredge mineral treatment plants have, for the most part, been superseded by jig treatment plants which recover finer sizes of valuable mineral and where the cleanup process is continuous. On large capacity dredges gravity distribution of screen undersize containing the valuable fraction of mineral to a multitude of jigs is difficult. There is an economic tradeoff between the capital cost of elevating material high enough to enable good splitting and distribution under the force of gravity, and the operating cost of pumping to distribute material to the jigs or other processing machinery.

Most modern bucketline dredges in tin, gold and diamonds use the assistance of pumps to distribute the washed and classified material rather than elevating it during the digging process and relying entirely on gravity for distribution. Recently constructed or modified dredges use pumps and relatively small splitters to distribute the classified material to a small number of relatively large circular jigs.

TAILINGS DISPOSAL

Dredge tailings in tin, gold and diamonds typically swell about 30 percent over their bank measure because the bank material is both completely broken up and classified with at least two size fractions disposed of separately. Where there is a high proportion of coarse oversize as in gold dredging with spud dredges, some of the oversize tailing is deposited at the stern of the dredge to five a firm seating for the spud. The balance of the coarse oversize is elevated and transported astern by a stacker conveyor suspended from the stern of the dredge.

The finer material resulting from classification aboard the dredge goes to the mineral treatment plant and ultimately to the tail slices to be send as far astern as the elevation of the treatment plant will permit. In tropical tin dredging where the proportion of coarse material is usually small, belt conveyor stackers are not generally used. However the fact of swell requires that some of this material be transported far enough astern and elevated far enough above water level to be disposed of so as not to fill the dredge pond with tailings. Sand pumps with sand pipelines astern can be used for such transport. Alternatively, stackers supporting pipelines ending in large elevated cyclones leave windrows of sand tailing above water level well aft of the dredge while returning the cyclone's overflow water to the dredge pond.

A third tailing is slimes, the clay, silt and fine organic material which settles to the bottom of the dredge pond. If this material is allowed to remain in a shallow dredge pond its solids first interfere with mineral recovery by floating waterlogged vegetation to the level where this material blocks pump intakes. This problem is quickly followed by an increase in the specific gravity of process water causing recovery in the gravity processing machinery to decrease. Finally these clayey solids can become thick enough to trap the dredge when the bucket line is idle for long enough to allow the slimes to settle. Inland dredges must pump the deeper layers of slimes out of their dredge pond and into containment areas, usually recirculating decanted water back to the dredge pond.

Where the ground for dredge mining has contained a large proportion of oversized rock, in the past this material has stacked astern on top of sand tailings. While this washed rock has in some places become a valuable gravel resource often worth considerably more than the mineral which was mined to place it there, undisturbed windrows of washed rock from placer mining will not support vegetation until they weather, usually decades from the time that they were put down as tailings. If it is necessary that an area to be dredged is to be restored to its original condition it must be remembered that unless some of the material can be removed and placed somewhere else the ground elevation of the mined area will be raised appreciably due to swell.

The other problems of land restoration can be solved at lesser cost if the dredge is **designed** with land restoration in mind. Tailings sand can be cut from tail sluices, dewatered and mixed with oversize from screens for discharge on the stacker. The stacker discharge can be made directable, to some minor degree, without danger of capsizing and stacker discharge can be directed to some further degree with high pressure water or slimes and the already essential slimes decantation areas can be placed on top of tailings to restore silt to the surface. All of this involves advance knowledge and planning, and increased capital and operating costs but it can be done without doubling the cost of mining and without major change in the mining method, **if it is planned in advance of construction**.

CAPITAL AND OPERATING COSTS

At a very rough approximation, the first cost of a mining dredge with 20 cubic foot buckets (in 1991) would be about US$ 25 million, without transportation costs. Dredge mining operating cost for such a dredge, adequately engineered and funded, as an approximation, varies from $1.50 to $0.50 per bank measured cubic yard of material mined, without land restoration considered. Some lower costs are claimed and other higher costs are known and not boasted of. A bucketline dredge can mine a coarse, free running, river gravel at very low cost. The total annual cost of running a specific bucketline dredge doesn't vary much with the yardage it digs, so the success or failure of a bucket line dredging venture depends very much on the yardage that the dredge can be made to treat on a consistent basis. this in turn relies on the accuracy with which the ore reserve has been evaluated so that the dredge is suited to the conditions that it will meet. A dredge must be designed to cope with every condition that it will meet; a single dredge mine is like a single stope mine with one drill and no stockpile: when the dredge is idle there is no income.

Chapter 6

ALTERNATIVE DREDGING METHODS

by Douglas J. Piper
Vice President
Ellicott Machine Corporation
Baltimore, Maryland

On reflection of certain overall circumstances which have pervaded the gold mining industry, it appears useful to point out some specifics which the writer hopes will improve the potential for success in the application of hydraulic dredges to gold mining activity.

I will, prior to entering into that discussion, comment briefly on rule of thumb for the application of the three types of equipment. The three are bucket line, bucket wheel and drag line dredges.

From a standpoint of high volume movement over a short distance, between 50 and 300 feet, the drag line dredge is historically the most cost effective type of machine to be utilized if bulk excavation is the single driving consideration. It is typical for this sort of equipment to operate in the 10 to 30 cents per cubic yard range, however, cost of operation is not the sole consideration. Typically the drag line mixes and remixes the material which it excavates. In the event that gold recovery is the object of the exercise, a strong potential for loss of the valued mineral exists. Additionally, drag lines do a poor job of bottom cleaning, particularly if there is a ridged or creviced bottom which must be cleaned and in some cases partially removed to assure recovery of the highest percent of valuable ore.

The bucket line dredge has, until recently, been considered an extremely cost effective device. However, the high cost of manufacture of this type of equipment has mitigated against its use in the majority of modern placer mining activities. Operating costs for this equipment run from as little as a few cents per cubic yard to as high as two to three dollars per cubic yard and higher under unusually difficult circumstances. Notwithstanding the cost of operation, the bucket line is an extremely effective piece of machinery and virtually the only satisfactory machine to be utilized in areas where high concentration of gravel and boulders are encountered. Generally speaking, in recent times the majority of new applications of the bucket line have involved the use of second hand machinery purchased at much less that new cost. It is the writers experience that bucket line dredges cost on an average four times as much money per unit volume of production as an equivalent hydraulic type dredge. Despite this, in many cases they are the only type of machine to be considered.

Bucket wheel dredges, a recent innovation in the placer mining industry, have demonstrated excellent capabilities and cost of operation vary from as low as 30 cents per cubic yard to as high as $2.30 per cubic yard. They produce excellent results in clays, sands, and relatively uniformly graded gravels which constitute less than 20 percent of the total through put of the dredge. Their ability to crush and provide a preliminary wash for the ore prior to its delivery to the process system make them extremely attractive to alluvial miners. The bucket wheel's principal shortcoming is it's inability to handle large stones (stones larger than the pumping capability of the dredge pump) and the very high wear rates which they tend to incur if heavy concentrations of gravel are encountered. It is conceivable that wear rates as high as one to two dollars per cubic yard can be incurred under very severe conditions. In the case of projects where very high values are encountered, this wear problem is not a serious consideration, however, such high value projects are very seldom encountered. The machines produce good bottom clean- up and have demonstrated great effectiveness in clay- bound tin mining activities.

This brings me to the comments I wish to address regarding the application of bucket line and bucket wheel dredges in recent years and the difficulties which have been encountered., I will address this on a broad basis in an effort to avoid naming names and specific organizations.

During the past twenty five years and particularly in the last decade the writer has had an opportunity to observe a wide range of activity in the alluvial and placer mining arena. It may be summarized as much smoke and occasional fire.

Two areas of exception have provided outstanding achievement and are the tin and mineral sands business where excellent success has been derived. Large scale gold mining activity has however been much less than successful.

There have been two principal sources of difficulty from which this the negative results have sprung:

1. Too much optimism with too little hard analysis and
2. Improper selection of equipment.

Reverting to the tin and mineral sands business - the greatly increased demand for mineral sands-derived heavy minerals and the development of very rich tin alluvials in Brazil have made these two areas subject to great success and very satisfactory application of the bucket wheel technology. The technology became available through development of the hydraulic dredge equipped with the underwater bucket wheel excavator. This new equipment has made possible the continuous low cost production of very large volumes of both heavy mineral sands and cassiterite.

This new type equipment makes possible high volume production, excellent size reduction of ore being mined and preliminary washing prior to presentation to the treatment plant. These factors have provided the dredge mining community with new and very effective technology and production capabilities.

Notwithstanding the new technology in the gold mining arena, we find a different situation. In principal all of the benefits which are available to the mineral sands and tin mining groups are available to the alluvial gold industry. The application of this type of equipment however has not been successful when attempted.

In six recent applications of the bucket wheel technology to gold mining, all failed. Economics, poor ore body analysis, and improper equipment selections defeated the operations. A seventh application was subject to the vagaries of national politics and never reached production and in fact was expropriated by the national government concerned.

The six bucket wheels which reached production status were all plagued with the same problem. Too little throughput and values which would not sustain profitable operation even if full predicted ore throughput had been achieved. In all cases our company had advised the mining organizations prior to equipment purchase that the projects were marginal at best and in some cases impossible to make profitable.

In three of the six projects mentioned, Ellicott supplied equipment. In two others we quoted equipment larger and heavier than that purchased and in the sixth we advised the owner that there was little if any likelihood that the project would break even, let alone make money. On another very large project we recommended a very large used bucket line dredge in preference to a hydraulic dredge due to the presence of significant quantities of large gravel and boulders. This project also has failed.

The principal difficulty with each of these projects was too much optimism regarding recovery of gold from marginal ore and selection of equipment which was not large enough to do the work.

In the case of the hydraulic dredge selection, the miners knowingly selected machinery which was too close to its maximum rating to be effective over the long term.

It is not the business of manufacturers to criticize the performance of a particular segment of any industry. However, when their product is involved in the activity on a regular basis it seems prudent to comment on some of the salient reasons for problems.

The dominant problem has been too much optimism untempered with a view of the past and the history of the gold mining as viewed from financial results.

The writer has, after considerable analysis, set a minimum value of $3.00 demonstrated recovery per ton of ore processed as the absolute minimum acceptable level at which modern dredging machinery and process equipment can be applied with a potential for profit. Below this, logistics, maintenance, and capital costs cannot be serviced with any chance of profitable operation.

All too often old rules of thumb have been applied to a modern economic analysis and they have failed. Additionally, promises from various manufacturers have generated unrealistic expectations on the part of would be miners and the results have been disastrous. Promises of nearly total recovery of fine gold or gold trapped in a clay matrix have consistently led to failure.

Selection of equipment which is rated at or near the required continuous throughput of the plant is also a certain formula for difficulty. In short, it would be well for any group or individual engaged in the consideration of alluvial mining to take a very hard look at the entire picture including equipment selection, demonstrated recovery capability of equipment and the specific project under consideration and finally a thorough understanding of the logistic and support functions of the specific project. In the recent past these items have been seriously ignored by many hopeful gold miners and the results have been predictable -- disaster.

Chapter 7

PLACER PROCESSING

by Louis W. Cope
International Placer Consultant
Denver, Colorado

INTRODUCTION

Placer material, regardless of the value of gold, tin, titanium, diamonds or other value mineral it contains, remains worthless until the valuable constituent is separated from the bulk of the material. This separation is no mean task when considering that a gold placer containing 0.01 to 0.03 oz/yd^3 of gold, which is the range of grade of typical placer deposits; yes, 1:3,000,000! It is a miracle that gold placer mining has ever been profitable.

Added to the situation of minute quantities of gold are several other problems, such as the gold often being very finely divided, the association of boulders and clay, and running and/or ground water. As each placer deposit is individual and distinctive, many factors have to be considered in planning the method for gold recovery from the alluvial deposit. Previous chapters have shown that in the exploration and sampling stage, some knowledge of the amount of each size of coarse and fine material which occurs in the deposit as well as the size distribution of the discrete gold particles is determined.

Unless informed to the contrary, as a general rule of thumb, plant designers will consider that 40 percent of the material will pass through a $^5/_8$-inch screen or punched plate hole. Also that there may be an occasional nugget and normally considerable amounts of fine gold. The designer must be informed as to the amount and stickiness of the clay, and the relative amount of black sand.

The stages of processing consist of washing, screening, rough separation and upgrading. This chapter considers only gold recovery, although the process is similar for recovery of other valuable minerals.

FEEDING

Grizzly

In most placer processing machines, a feed hopper receives the excavated muck. Most of these have a grizzly, which is a set of bars or rails with a pre-determined opening between them, for removing boulders. It is best if the oversize rock can be washed prior to being discarded, with the wash water remaining in the circuit.

Static Grizzly: When the subsequent machine is a trommel, the grizzly opening is generally in the 6-inch to 12-inch range, depending on the trommel diameter. If a vibrating screen follows the hopper, the grizzly opening is usually in the 2-inch to 4-inch range. Water spray bars can be set to wash the boulders prior to their discharge.

Vibrating Grizzly: There is a machine on the market which receives feed to a hopper that is connected to grizzly rails, all of which vibrate, and has water spray bars. Grizzly openings on these machines are normally set at less than 4-inches. Vibrating grizzlys are commonly used in gravel plants, but can be used with gold placers if there is a minimum of clay.

Ore Feeder

As mentioned, feed can be put on a hopper-grizzly, with the ore fed to the next stage by being slurried and flowing by gravity. An important item to bear in mind at this stage is that the ore should be treated as soon after being dug as possible. As it is usually moist or wet when excavated, it should be washed and slurried before the clay dries and hardens.

Static Feeding: Muck which is dropped into a hopper or sluice box depends on the water to slurry and move it. With proper adjustment of the water flow, the ore will feed to the next stage at a fairly constant rate.

Positive Feeding: Plate, reciprocating, vibrating and conveyor feeding fall into this category. The reason for using positive feeding is to control the amount of ore fed, and to have consistent versus surges of feed. However, these feeding machines complicate the installation, are another item to maintain, and consume power.

SCRUBBING AND SIZING

This section will primarily discuss trommels. A trommel is shown diagrammatically in Figure 6. It consists of a cylinder with the axis inclined slightly, usually from 0.5 inch to 1.5 inch per foot of length. The length is several times the diameter. A spray-bar for wash water is in the center of and independent of the drum. The trommel is supported by rollers, and rotates at 20 to 30 percent of critical speed. This relates to 8 to 12 revolutions per minute for a 5-foot diameter trommel.

Scrubbing

This step overlaps the washing stage, which occurs with most processes of treatment. Scrubbing is normally done in the first portion of the trommel, where blank plates are used, along with lifter bars, to agitate and mix the muck, and retarder rings, to slow advance of the slurry. Water enters with the ore from the feeder or hopper, and more water is added from the spray-bar. The larger rocks which were allowed to pass through the grizzly also aid in agitating the slurry and in breaking up lightly cemented ore.

Sizing

The middle and lower end of a trommel has punched plate or screen to allow the water, slime, sand and gravel to be separated from the large gravel and cobble. A trommel will often have two size holes or screen, with the smaller holes first. These should not be smaller than $^3/_8$ inch. Holes smaller than this size will get plugged too easily with vegetable matter which comes in with the feed. The larger holes are usually $^5/_8$ or $^3/_4$ inch. The reason for these larger sizes will be explained in the paragraphs on sluices and jigs.

Bear in mind that a hole of any diameter will, in effect, pass a maximum size rock particle of about $^1/_8$ or $^1/_{16}$ inch diameter smaller than the hole; *i.e.*, a $^1/_4$ inch particle through a $^3/_8$ inch hole. In a trommel, the holes in a punched plate should be made in flat metal sheets, then rolled for the trommel cylinder. This is to give slightly trapezoidal holes which are self-cleaning, hence less apt to become plugged by small rocks.

Screens have been used for sizing and washing placer material. When used they are usually multi-deck, and have spray-bars. Screens are much less effective with clayey ores, and require higher head room than a trommel of equal ore throughput.

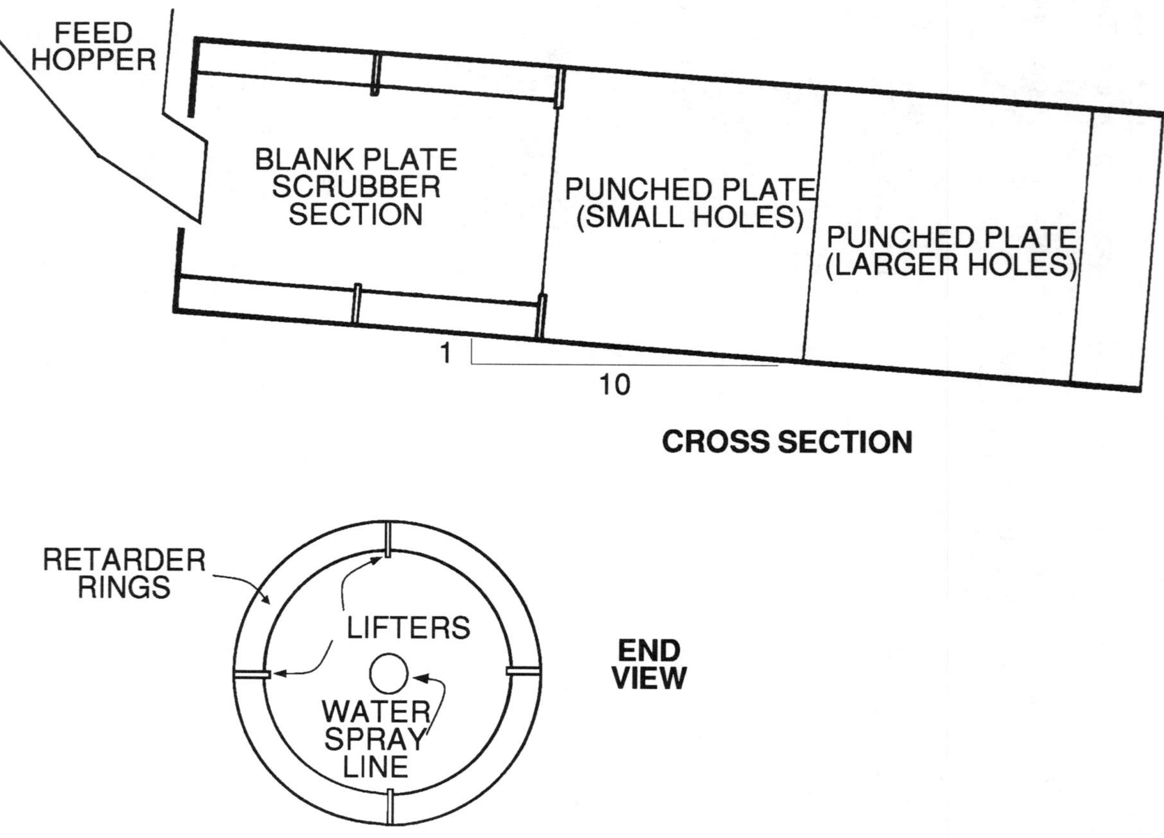

Fig. 6. Components of a trommel used for scrubbing and screening.

GOLD TRAPS

In the normal placer deposit, the largest nuggets are small enough to pass through the screening stage, to be saved in the first concentrating machine. However, some operators will want a point where the nuggets will be collected, called a trap, in the stream where the screen undersize flows or at the inlet of the concentrating process.

It is an unusual alluvial deposit which will have the rare nugget larger than the holes in the screening step. In some instances, a nugget trap is put into the chute where the feed enters the trommel. The disadvantage to this is that rocks or pieces of wood will be trapped and can obstruct flow of the muck to the trommel. Consider that the saving of one nugget every once in a great while will not make an uneconomic mine profitable. I did see at a property in Alaska known for its occurrence of nuggets a 2-foot section of the trommel with larger holes to feed a special sluice just for nuggets. In addition, there was a metal detector mounted on the screen oversize conveyor to waste.

CONCENTRATING MACHINES

There are any number of machines and methods used for recovery of gold from placer deposits. Each will have its own advantages, disadvantages and field of use. Because each placer deposit has its distinctive characteristics, several of the commonest machines will be described here. Even then, some are commonly used while others are rarely used. In this discussion, the methods will be divided into passive, reciprocating and centrifugal machines.

Passive

Sluice: Stationary sluice boxes with any of a variety of types of riffles are the historic method for recovering placer gold in small scale. Several variables can be designed into a sluice system. Size of rock and gold will be the factors which affect the amount of water, sluice width and length, slope and type of riffles. The slope of a sluice handling a typical minus $^{1}/_{2}$ inch material with one part of solids to three or four parts of water is $1^{1}/_{4}$ inches per foot of sluice length; *i.e.*, approximately 1 in 10.

The commonest type of riffles used are called "Hungarian" riffles and are shown in Figure 7. These can be made of wood, but are usually of angle iron. Sets of Hungarian riffles are formed by welding several angles at a set distance apart to two runners on the ends of the angles. This facilitates sluice clean-up, as lengths of riffles, usually 4 feet, are lifted out. Fine gold is best saved when in the presence of larger gravel in the sluice feed. The gravel gives small spaces where the fine gold particles can settle and be trapped.

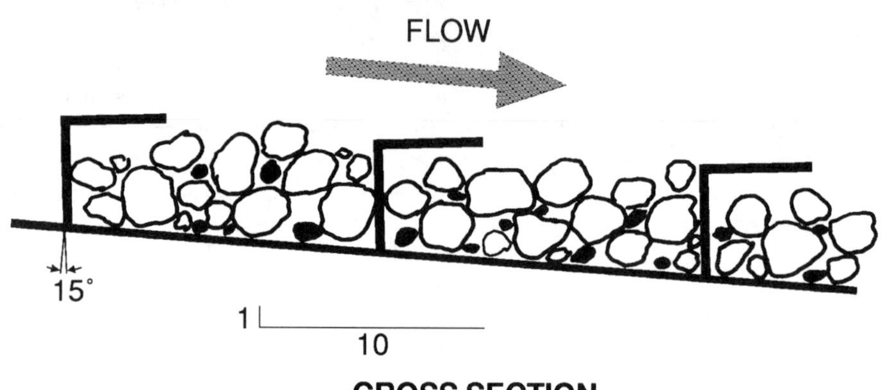

Fig. 7. Sluice showing angle iron Hungarian riffles. Large size waste gravel assists in the recovery of fine gold.

Fabric: Indoor-outdoor carpeting, Astroturf or gunny sacks placed under riffles will facilitate the collection of fine gold. Another method is to use one of the fabrics and put expanded metal over it. In this, the long dimension of the expanded metal diamond is placed transverse to the direction of flow. Also, the "tilt" of the metal should be so the metal overhang is toward the downstream end of the sluice.

The mention of the "Golden Fleece" in Greek mythology could have been an ancient practice of a fabric use for collection of placer gold. The natural oil in a sheepskin would aid the collection of fine gold.

Amalgamation: In early-day sluicing of ore with fine gold, mercury was a major collector. The alluvial ore would be fed to a sluice with some mercury held by the riffles at the upper end of the sluice.

Thin Film: When the slurry of a screened feed of fine material is fed over a smooth surface, the heavier particles will migrate to the bottom of the film of water. In so doing, they will drag on the surface and slow down, while the overlying water with the lighter waste particles will flow faster. This phenomenum is used in cones and spirals and is shown in Figure 8.

Cones -- In cones, the feed slurry is actually fed to a smooth surfaced conical shape. Near the lower end of the cone is a small annular ring opening. The gold drops through this while the momentum of the faster flowing waste-carrying slurry carries it over and past the ring opening.

Spirals -- The same thin flow concept to concentrate gold and other heavy particles is used in spirals. But in this machine, the slurry race is designed and manufactured so that the drag of the heavy particles on the spiral surface carries them toward the center, while the velocity of the slurry carrying the waste takes it toward the outside of the race. It is then a simple matter to have a divider or ports to cut the two products into two separate streams.

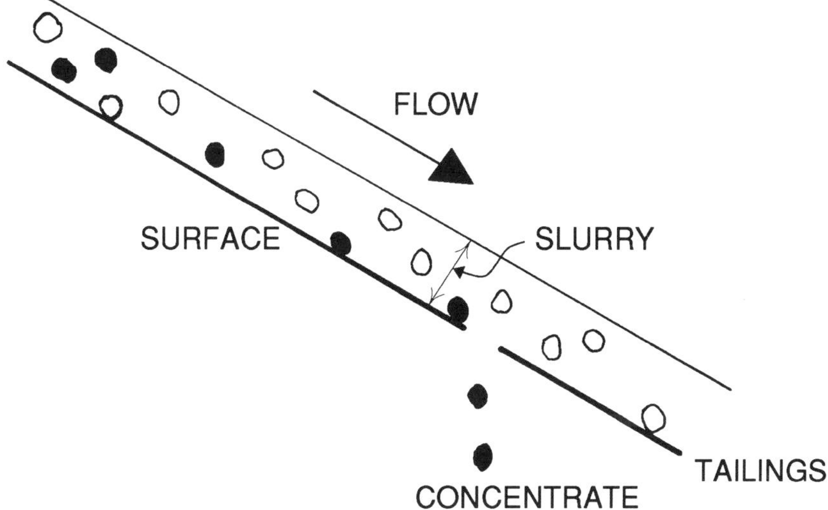

Fig. 8. How a thin slurry film concentrates gold. This is the basis for spirals and cones.

Reciprocating

Jig: A standard item of gold recovery on most major dredges is the jig. This machine, shown diagrammatically in Figure 9, is named for the jumping or jerking motion built into it. The feed slurry flows across the machine above a screen or punched plate usually of $^{1}/_{8}$th inch mesh. Below the screen is a conical lower compartment into which a pulsing motion is imparted, and water injected. As the feed flows across the upper compartment, each pulse raises it. Between pulses, the heavier particles settle downward faster than the lighter ones thereby separating the gold and other heavy particles form the waste. There is a bed of small steel balls called "ragging" on the screen which further facilitates the separation. Gold or heavy mineral particles too large to pass through the screen will collect on the screen.

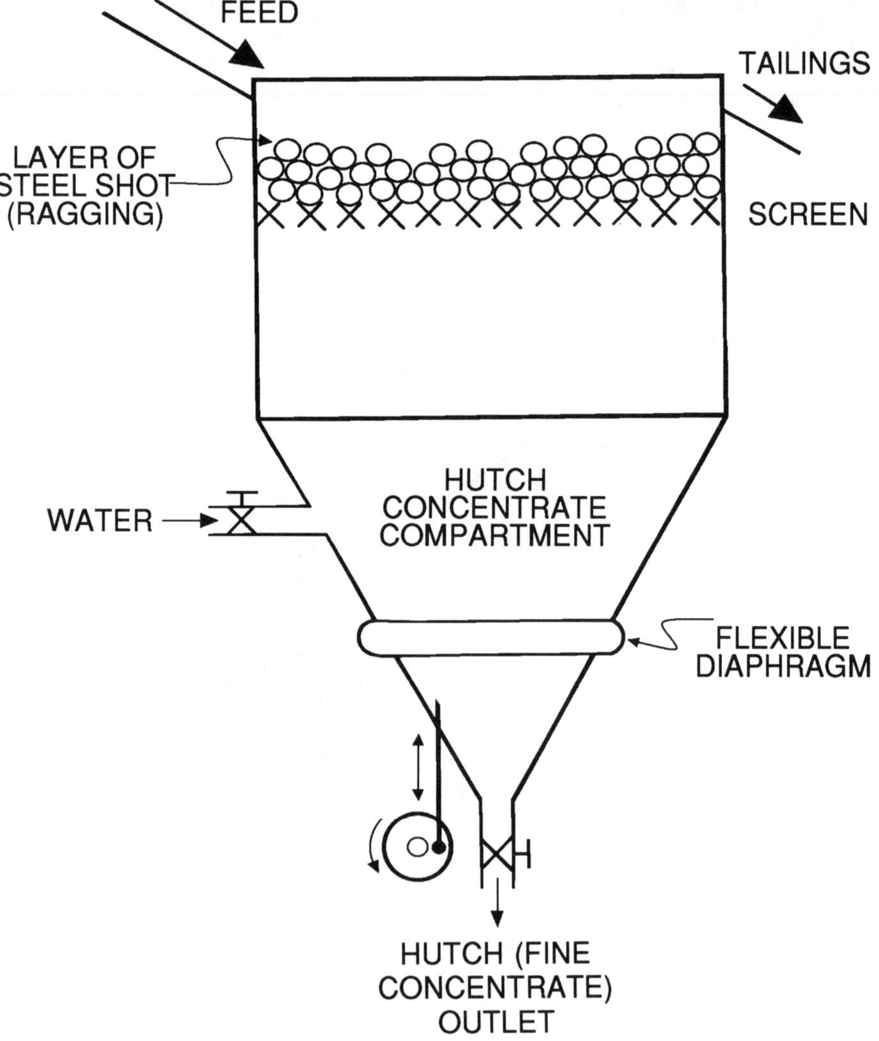

Fig. 9. Components of a jig.

Vibrating Table: A flat deck which reciprocates in a horizontal plane also separates the heavy from the light-weight particles. A diagram of a vibrating table is shown in Figure 10. The table deck has riffles in the long direction of the table. These are about $^3/_8$ of an inch high at one end and decrease in height to about $^1/_{16}$ inch at the concentrate end of the table. The table motion is a "quick-return" type; *i.e.*, the movement toward the concentrate end is slow enough to carry the slurry with it. However, the return motion is so fast that the table is pulled out from under the slurry, thereby causing the slurry to migrate toward the end of the table. As it gets to the lower height portion of the riffles, the lighter particles wash over them toward the tailings side of the table, while the heavier particles continue to be moved toward the concentrate end.

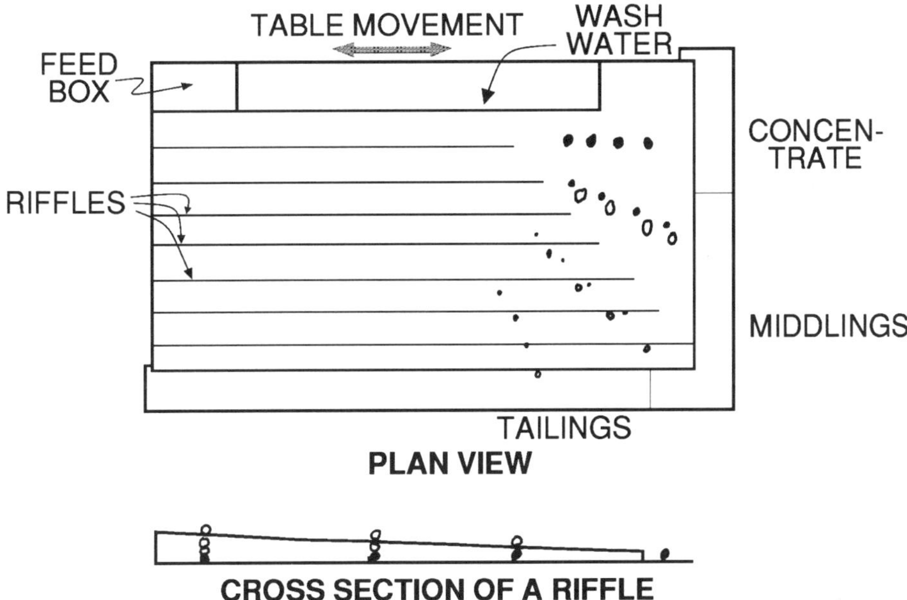

Fig. 10. How a vibrating table separates mineral components.

Panning: The motion in hand-panning is more or less a revolving movement with a little eccentric or jerking superimposed. The latter causes the gold and other heavies to settle to the bottom while the swirling imparted by the revolving motion washes the light sand off the surface. Panning is an art which has to be taught by demonstration as opposed to description by the written word.

Centrifugal

Bowls: Most major dredges in their 90-year history have used centrifugal bowls as a first stage of clean-up of the rougher concentrates. These machines, usually no more than 3-feet in diameter, are driven from the bottom, revolving around a vertical axis to several times the force of gravity. The interior of the bowls have concentric circular riffles or baffles. Feed is to the center of the bowl. Volume of the feed and configuration of the bowl sides force the slurry over the riffles to the top where it overflows to waste. Meanwhile, the heavier particles settle to the outside and are trapped by the riffles.

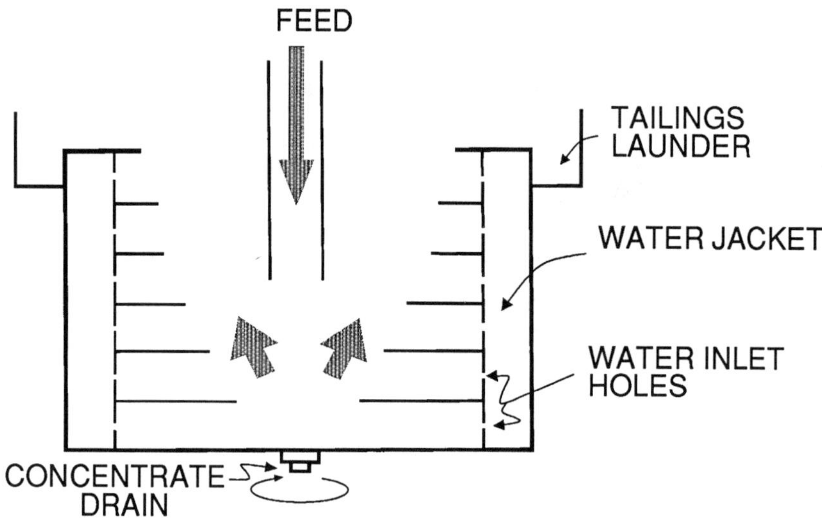

Fig. 11. Components of a centrifugal bowl showing the water jacket.

In recent years, a second generation of this machine has come on the market. A diagram of the device is shown in Figure 11. It has a water jacket and injects water into the accumulated concentrate held by the riffles or baffles. This action keeps the concentrate loose, like quicksand, while upgrading it by washing out some of the sand.

CLEAN-UP

After a rough concentrate produced by any of the machines described above, a more concentrated gold product has to be made, which can be accomplished in one or more steps. This is called "clean-up". This upgrading can be either continuous or batch processing. Often times the same machine, perhaps a smaller model, will be used in the clean-up as is common practice with sluices and jigs.

However, as the bulk of the concentrate is reduced in size and weight, other machines are commonly used, such as tables, bowls and spirals. Finally, small dish-like machines with spiral riffles, hand work, amalgamation, fire refining, or other methods are used to produce the gold product.

CONSIDERATIONS

Each placer deposit will be distinctive in its characteristics and each will vary from place to place within the same deposit. the relative quantity as well as the shape of boulders, cobble, gravel, sand and clay will vary. Likewise, the quantity, size and shape of the gold particles will vary. Then there is the possibility of natural cementing of the material to varying degrees. All these factors affect gold recovery and should, therefore, be taken into consideration when designing a plant.

There are also mechanical factors to consider. This includes trade-offs to be able to treat more ore per unit time, although another machine might give a little better recovery. Items to consider in this light include sturdiness of the machine, complication, power consumption, ease of repair, "forgiveness" as to changes in feed rate and/or particle size, operation ease, and size of feed acceptable, space and headroom required, and cost, to name a few. Table 3 summarizes some of the advantages and disadvantages of the various processing methods discussed.

Table 3. Process Methods, Advantages and Disadvantages

Method	Advantages	Disadvantages
Sluice	• Simple • Inexpensive • Construction materials available everywhere • No power required	• Reduced gold recovery • Time-consuming to clean-up • Low capacity
Fabric	• Simple • Inexpensive • No power required • Improved fine gold recovery • Best used in conjuction with a normal sluice	• Low capacity
Amalgamation	• Simple • Inexpensive • No power required • Recovers very fine gold • Best used for final clean-up	• Low capacity • Cost of make-up of lost mercury • Legal and moral constraints • Not all gold will amalgamate
Thin Film	• No moving parts in separating machine • Recovers fine gold	• Requires screening to fine size • High headroom required • Pump and piping wear
Jig	• Feed-size forgiving • Feed-rate forgiving • Requires little operator attention • Recovers fine gold	• Complicated machine • Several adjustments

Table 3. Process Methods (continued)

Method	Advantages	Disadvantages
Vibrating Table	• Can produce high grade gold concentrate • Takes jig concentrate as produced • Best for clean-up • Separation clearly visible	• Requires screening to fine size • Requires considerable operator attention
Bowl	• Recovers fine gold • Requires little operator attention • Long history of use in clean-up • Longer run gives better concentrate	• Requires screening to relatively fine size • Has to be stopped to be cleaned up • Requires absolutely clean water (for water injection models)
Pan	• Minimum equipment required • Simple • Inexpensive • Often used in last stage of clean-up • Best used for prospecting	• Very low capacity • Hard work

RECLAMATION

The environmental aspects and reclamation of placer mining is a subject unto itself. It will suffice to say, here, that there are legal and moral obligations to keep these matters in mind. A properly designed mining and tailings disposal plan can do much, if not most, of the reclamation concurrent with mining.

IN CLOSING

As has been stated in this chapter, each deposit is distinctive in nature, so consideration of the characteristics of each placer occurrence must be made. I have told clients for years, "The nature of the muck tells you how to proceed, you'd better listen." I strongly recommend the use of time-proven industry-standard machines and methods instead of reinventing the wheel.

Good planning and simplicity of operation will increase profits for the operator.

Chapter 8

THE GOLD PLACER INDUSTRY, PRESENT AND FUTURE

by Mortimer J. Richardson
President
CONSOLIDATED PLACER DREDGING CO. (CPD)
Irvine, California

INTRODUCTION

In the context of this session on "Practical Placer Mining," a projection of the present into the future first assumes that there is a future for Placers. The second assumption is that there is the combination of adequate reserves and a willing, mining investment source to finance placers. This paper will point out the opportunities on both assumptions, illustrating my positive position on the future of gold placers. In order to focus this paper more effectively, I have chosen GOLD as the subject mineral. However, it is recognized that much of the technology for exploring and mining gold placers has been applied to other alluvial deposits; including platinum, diamond, tin, and various industrial minerals (which are often referred to as "placers" though this contradicts historical definitions).

Over the past 10 years I have presented several papers on the general subject of placers, most of which are listed under References. These have included topics such as exploration, evaluation, small-scale gold placer mining and decision-making, mining and processing equipment designs, and dredge-mining operations. In an attempt to identify the more promising, undeveloped placer potentials in the world I have usually mentioned the USSR and China but qualified them as being either inaccessible, or, politically difficult to obtain. However, with recent developments in the USSR, the obstacles appear to be somewhat diminished.

CURRENT OPERATIONS OF GOLD PLACERS

It is known that there are extensive workings and deposits of gold placers in the USSR and in China. Outside of those countries, the largest gold dredging operation is in Colombia; Mineros Antioquia SA, on the Nechi River with five large BL/M dredges. Other areas known to us to have gold mining dredges producing on a commercial scale (*i.e.*, 50,000 m^3/month or more), are as follows:

USA	California	(1)
	Alaska	(2)
South America	Colombia	(5)
	Peru	(1)
	Bolivia	(1)
	Brazil	(1)
Africa	Ghana	(4)
	Guinea	(1)
Indonesia	Kalimantan	(4)

My papers dated 1980 and 1982, contain photo illustrations and data on contemporary as well as earlier, gold dredges.

In addition to large scale gold dredging, there are literally thousands of small, suction/riffle gold mining "dredges" operating throughout South America, Africa, Southeast Asia, Indonesia, New Zealand, Canada and the U.S. Few of these qualify as commercial scale, mining operations by themselves though when added together in given areas, such as the Madeira River, Brazil, their production is considerable.

Typical of these operations are the small tethered barges with a 7.5 to 15 cm (3-6 inch) pump and engine, discharging onto an apron that tapers into riffles. The suction hose is either handled by a diver, working on the bottom of a river; tied to a pole that is pivoted to the bottom, or, simply hung from an "A" frame and worked by hand from the barge. Very little fine gold is recovered since it normally is carried away in the turbulence of the flume arrangement. I might add that the system described is the more "sophisticated" method of placer mining while tens of thousands of local miners will be content with a crude shovel, pitting to depths of one meter, sluicing and panning, working the terraces along rivers and ponds. These qualify for the "garimpeiro" style of mining sometimes characterized as "illegal miners."

USSR Gold Placers

Under the spirit of "Glasnost," great strides are being made to open up the natural resources of Russia. The article by Kvint (1990) shows an aerial photo of a large, BL/M dredge ostensibly mining diamonds. It is clearly of the California-type design and possibly a copy of the Yuba dredges sold to Russia; in 1925 four, 13½ ft^3 and one, 7 ft^3; in 1947 seven, 7½ ft^3, all of the standard Yuba BL/M or California-type. The 1925 dredges were for mining platinum while the 1947 orders were all for gold placers. Kvint urges Western companies to come into the mineral rich areas of the Russian Far East, to make joint ventures or complete investments of 100 percent, and that both labor and support facilities will be abundant.

A more revealing presentation was given along with slides by Bogdanov in 1990 during the Minerals Symposium sponsored by the Alaska Miners Association. He indicates that gold placers are mined only about 15 percent by dredges; the rest by crude methods of portable washing plants, aided by dozers and earth haulers of various types. Others are mined underground in large caverns, in the permafrost. Stripping of the frozen ground is done by dozers to permit natural thawing before mining either by dozers, frontend loaders, or dredges. He states that about 70 percent of the placers in Russia are in permafrost. The underground mining permits 12 months per year operation while surface mining is limited to the milder periods of 5-6 months per year. They have also been practicing a great deal of blasting of permafrost but Bogdanov considers this inefficient and has called for a significant reduction of that method to be replaced by other methods, preferably natural thawing. His discussions indicated significant differences and complaints about central government interferences, or neglect of placers. As a result he states that about 70 percent of the placers are still mined using sluices, or, riffles for recovery of gold; thus losing the fine gold. Because of the cheap labor the government shows no interest in cost of operation or pleas for higher production dredges, use of jigs, and other modernizations. He states that the dredges are too small and should be increased in size for greater production. Their dredges are limited to 15-20 m (49-66 ft.) digging depth and need to be increased. Most mining for gold placers is being done as "70-80 years ago." His further plea is for more exploration and dredge technology. Bogdanov is now 77 years old and has operated mainly in R&D of gold placer technology for the past 37 years. He believes that further exploration will open up new deposits of gold placers.

The article in *E&MJ*, confirms the statements of Bogdanov, showing a photo of a basic washing plant. Bogdanov illustrated his talk with several slides showing the same equipment with variations of conveyors and trommels but all feeding into sluices for recovery.

Using the low cost labor of both Russia and its trading partners, this may be a window of opportunity for the development and mining of large gold placer deposits. Conditions will be harsh, but the rewards could be of a large magnitude.

Chinese Gold Placers

In the past 15 years, China has opened up to outside procurement of BL/M dredges for gold mining. Some European companies have sold their designs of BL/M dredges and provided technical assistance, fabricating the dredges in Chinese shipyards in Shanghai. Those dredges are reported to be mining at various locations in China.

As reported by Cleaveland, (1973), he and his partner Frank Adams sold one 8 ft^3 BL/M (used) Yuba-type, dredge to a Japanese company and shipped it from California to Manchuria, in 1939. Cleaveland found some of the parts of the dredge near a railroad in 1947, when touring Manchuria. Cleaveland's book (1973) is instructional on how to operate a placer dredging company under difficult, war-time conditions. In about 1979, I met with a representative of a Japanese company who stated that they bought Cleaveland's dredge in 1939 and made several duplicates of it. Following on to that, an inspection was made by one of our engineers of gold placer dredging operations in China, where he saw ten 8 ft^3, Yuba-type dredges.

Our records show the following BL/M dredges purchased from Yuba Manufacturing for shipment to China in early days.

1928	1 - 10 ft^3 buckets, for gold placers.
1939	2 - 7 ft^3 buckets, for gold placers.

In other areas of China, however, many gold placer deposits are being worked by hand methods. There are areas of placer potential yet to be developed but require extensive exploration programs to evaluate and prove them.

African Gold Placers

The African continent has opportunities for considerable development in gold placers. Private companies are developing new mines; in Guinea, Mali, Zaire, Sierra Leone, Ghana, Zimbabwe, and exploration is going on in other regions as well. The executive of a leading mining company remarked recently that the African Continent is more favorable for mine development now than in any time since before WWII.

While the political situation may be more open in African countries generally than in the USSR and China, the difficulties of logistics, infrastructure and communications are similar in magnitude. In some cases these problems become the dominant ones for the mining company and should not be underestimated in terms of planning and budgeting from the early stages. Likewise, however, evaluation of reserves through to design of proper mining equipment, become even more critical due to the "too late to turn back" phenomenon. Once you are in, it is an expensive exercise to make corrections to say nothing of admissions of costly mistakes.

FINANCING OF GOLD PLACER MINES

While financing of mines is a generic subject that shows no particular differentiation among types of minerals, some fundamental requirements must be met. The basic categories of gold mine financing can be summarized as follows:

Conventional production loan.
Project loan.
Commodity loan; gold futures.
Equity financing; selling of shares.

For a particularly good and inclusive paper on this subject, see Burton (1990).

In all of the above cases, financing must be substantiated by a feasibility ftudy, either performed by a third party to the mine owner, or verified by an outside authority or consultant. Where the mine is located in developing or politically sensitive countries such as would be the case in the USSR, China and a number of third world countries, the participation of the World Bank or a government agency may be necessary to guarantee or participate in the financing. Some of the essential ingredients for such a study are discussed below.

Evaluation

Evaluation should be performed using experienced placer mining engineers, employing methods that have an established validity. The methods of evaluation need to be based on time-tested procedures that have been verified with actual, successful gold production in commercial-scale quantities. There must be minimal influence in the study from manufacturers of equipment that might bias the results in their favor. This is discussed in detail in my paper dated 1987. However, a more exhaustive work can be found in our 1984 "Handbook of Alluvial Gold Evaluation Methods."

Proven Reserves

The Feasibility Study should be based on Proven Reserves, determined by standard methods. Drilling and evaluation by Churn Drill has been substantiated by the production of billions of cubic meters of dredged material, by leading gold dredge mining firms in years past. Those leading companies would include:

Alaska Gold Co., (USSM&R) Alaska.
Bulolo Gold Dredging Co./Placer Development Corp. (former affiliate of CPD), New Guinea.
Natomas Corp., California and Peru.
Pato Consolidated Gold Dredging Ltd. (former parent of CPD), Colombia.
Yuba Consolidated Gold Fields Corp., California, Alaska, Canada.

Bulk Sampling includes shafting, trenching, and pitting which are all methods of checking drill results, or, for initial exploration prior to drilling. For that reason, they should not be the only method of establishing Proven Reserves because of their limitations to produce a reliable average over a large area. In other words, large bulk samples tend to over-emphasize the tenor of the sampled locations.

When establishing Proven Reserves of gold placers based on the foregoing evaluation methods, there is an important criterion that should be used. In the past few years, there has been a tendency to quantify Proven Reserves in terms of the "pay" zone only. Thus there may be barren or clay-bound strata from the surface to some point above that zone, which is classified as "overburden." By stating the tenor in terms of the pay zone only, a distortion of values is created. The qualification given (*i.e.* "only includes the pay zone") is often obscured in the feasibility study so that the financial institutions and investors are misled into believing that this is the overall tenor. The fact is that **all** the material must be mined in order to recover the gold. For that reason tenor should be stated in terms of full dilution of material from surface to bedrock.

Equipment Design Study

An important function of the evaluation phase is to establish an understanding of the nature of the deposit to be mined; *i.e.*, clay, boulders, abrasiveness, organic matter, etc. The sizing, or, sieve analysis of *in situ* gold will influence the design of processing equipment and particularly in the final stage of recovery. Most gold placers are found to have their greatest concentration near bedrock. It is therefore essential that the dredging equipment be able to clean the bottom thoroughly and uniformly, so as not to require multiple passes over the same ground.

Unfortunately, in the past few years there have been some major failures in gold placer mines where attempts have been made to use untried designs of placer mining and recovery equipment. Somewhere in the process of the feasibility study, the requirements for tested and proven designs and methods were omitted. This subject will be dealt with in the following section discussing Placer Mining Technology.

Once the equipment selection has been made, data relating to the average production, energy consumption, and labor, is established from prior operating records. This historical data is used to estimate gold production and cash flow for the system.

My papers dated 1986 and 1987, were intended to set forth a basis for evaluating gold placers, though emphasizing "small-scale" mining. This was to give a basis for the minimum size mine decision but contains useful data on larger BL/M dredges. So often the mine development effort appears to have begun with a pre-conceived equipment configuration, attempting to make it fit; regardless of capital cost, efficiency and operating cost. The results are often the same; failure, after the stockholders and banks have lost their patience.

Cash Flow Analysis

The analyses of reserves, equipment design and historical operating records, produce the basis for determining cash flow. Once settling on an accepted gold price, the tenor of the deposit can be treated on either an average, or, progressing dredge course basis. This decision will depend upon how much variability in the tenor (Au mg/m^3), there may be in the deposit. The obvious aspect of this is that there must be a solid basis on which to project what amount of gold will in fact be produced; on a regular, monthly and yearly basis.

Operating experience with the equipment design can produce the basis for the estimated cost of operation, making allowance for revisions due to local conditions and updated costs for fuel, labor and outside services. When this is known, the bottom line is produced and will show if there is sufficient margin to repay capital costs, depreciation of currency and risk of investment.

This discussion is overly simplified but contains the essential elements on which to make the calculations and basis for an investment decision in the project.

PLACER MINING TECHNOLOGY

The general subject of Placer Mining Technology includes:

- Exploration and evaluation methods;
- Mining and processing equipment;
- Techniques of operation and production.

While some industries have experienced quantum jumps in technology over the past 40 years, placer mining has not had similar levels of financial investment or activity since WWII. Ancillary components such as electronics and electrical controls, plastics and other materials have offered the best opportunity for applications to placer mining. Developments in the offshore oil industry have likewise provided technology changes that are applicable to placer dredge mining. The ultimate test in all cases of new technology however must be whether the alternative will produce an improvement in recovery and cost reduction.

Exploration

While Reverse Circulation (RC) Hammer Drills have been used in evaluation of gold placers since the mid-1960's, there has been only limited verification by large-scale production. (In the case of tin there have been substantial verifications offshore Thailand). The principal substantiation with gold placers occurred in recent years offshore Nome, Alaska, with the BIMA dredge. Other types of drills including sonic or vibratory, have essentially no production verification that could be used in a well-qualified feasibility study for gold placer deposits. (This subject is treated at length in our 1984, Handbook of Alluvial Gold Evaluation Methods).

Our own experience with RC Hammer Drills (Becker), has been generally favorable though they need calibration with proven, churn drill systems. In 1981 in New Zealand, CPD conducted extensive drilling and verification with an RC drill, completing the tests with a 20 ft^3 BL/M dredge mining through the drilled grid. The results substantiated a revised adjustment factor for evaluation using the RC drill. The advantage of being faster (by a factor of 10), suggests that the RC drill should be used wherever possible. The major disadvantages are its weight and cost, limiting it to easily accessible areas and large deposits.

Using multiple drills such as the manual operated, Banka-type in low-cost labor areas can often be adequate and much cheaper. Drilling into deeper ground may make such a decision impractical, however. The dependable Keystone-type of churn drill, can still be found in some areas.

In the final analysis, the financial backer must be satisfied that the proven reserves are just that! based on substantial experience and qualification to make the assessment.

Dredge Mining Equipment

When evaluating a gold placer, finding that the average tenor of all material in the deposit may vary between 200-300 mg/m^3, you soon come to the conclusion that every particle of available gold needs to be recovered. This means that all material that is shown to be gold bearing in the smallest degree, should be processed if possible.

To date, there appears to be only one type of placer mining equipment that can accomplish a thorough cleaning of bedrock on a cost-effective basis. That is the bucket ladder mining dredge (BL/M). This is due to the function of digging on the catenary of the bucket line, dragging several buckets along the bottom as they approach the lower tumbler before reversing direction for dumping. In addition, the weight compensating effect of the continuous bucket line, reduces the total power required to mine or excavate. More effective methods and equipment are still to be proven.

Equally important is the treatment plant; to classify, concentrate and recover the gold on board the dredge. One of the most significant developments in the past 30 years, has been the "Cleaveland Circular Jig," which was originally patented in 1967. It has been applied extensively to tin, was developed for diamond placers in Brazil, then tested and installed on some of our former gold dredges in Colombia. In 1979, Cleaveland designed the second generation improvement together with CPD, called the "Mk II." This design makes high production possible combined with efficient concentration in a wide range of gangue materials (see my 1983 "Handbook of Mineral Jigs").

CONCLUSIONS

Gold placers have the advantage of lower capital cost to develop, but with attendant lower tenors. Placer mining engineers are scarce and great care has to be taken in evaluating, designing and operating a gold placer mine. Newer mining companies should seriously consider gold placers as part of their development program, recognizing that gold placers were the building blocks of many of the major international mining companies.

REFERENCES

Anon., 1990, "The Klondike -- Soviet Style," *Engineering and Mining Journal*, Nov.

Bogdanov, Y. I., 1990, "Placer Mining Technologies in the U.S.S.R.," Alaska Miners Association Minerals Symposium, Anchorage, AK.

Burton, A. K., 1990, "The Project Financing of Small to Medium Mines: A Banker's Perspective;" *Financing of Gold Mines Seminar*, Southern California Mining Section of SME, Irvine, CA.

Cleaveland, N., 1982, "Circular Jig," U.S. Patent No. 4,310,413.

Cleaveland, N., 1973, "BANG! BANG! IN AMPANG, Dredging Tin During Malayas Communist Emergency," Symcon Publishing, Irvine, CA.

Kvint, V., 1990, "Go East, Young Man," *Forbes Magazine*, New York, Nov., p. 234.

Richardson, M. J., 1988, "Finding & Developing Placer Gold Mines," *International Congress on Dredging Mining Systems*, Ellicott Machine Corp., Baltimore, MD.

Richardson, M. J., 1987, "Site Evaluation Criteria, Exploration and Development of Small-Scale Placer Gold Mines," *Small Mines Development in Precious Metals Conference*, SME, Reno, NV, p. 69.

Richardson, M. J., 1986, "Evaluation/Decision Process for Small-Scale Placer Gold Mining;" *Mining Magazine*, London, April, p. 312.

Richardson, M. J., 1985, "Placer Gold Mining; A Return to Basics," *World Dredging Magazine*, July, p. 12.

Richardson, M. J., 1984, *Handbook of Alluvial Gold Evaluation Methods*, Consolidated Placer Dredging, Irvine, California, 300 pp.

Richardson, M. J., 1983, *Handbook of Mineral Jigs*, Consolidated Placer Dredging, Irvine, CA, 200 pp.

Richardson, M. J., 1982, "South American Dredge Mining," World Dredging Magazine, Irvine, CA, July, p. 12.

Richardson, M. J., 1980, "Dredge Mining Placer Gold in the 1980s," *WODCON IX Proceedings*, Western Dredging Association, Arlington, VA, p. 809.

Chapter 9

PANEL DISCUSSION AND AUDIENCE PARTICIPATION

Panel Members: L. Cope, Moderator
J. Noble
D. Piper
M. Richardson
J. Wojcik

L. Cope

Welcome. I am pleased to see so many people interested enough to stay until this last session of the last afternoon of the convention.

Of the people on the panel here, everyone has done everything in placer mining, but we have tried to sectionalize it with each one of us taking a particular portion. Any one of us could probably answer any question. I am very pleased to see some well experienced people in the audience. To start the discussion, Joe, can you give us a brief comparison of pit and drill sampling?

J. Wojcik

They both have their place. If you expect your deposit is something that is going to have widespread fairly uniform grades, then the pits may be the way to go. If you are going to have something you have to trace for several miles down the river valley, you can't do it with pits and do a proper sampling job.

C. McLain

Joe, the end result of all of this is to go ahead and mine it if it is a deposit. Do you try to relate your drilling results to something in mining itself against the recovery, because the drilling in different types of ground will give you different kinds of recoveries, and mining in the end is what you are going to do? Do you try to relate this to your mining method?

J. Wojcik

Personally I don't. My preference is to record the best guess of what is really there and record the sediment conditions and the clay and boulder content. Let the engineers make the adjustments down the road. Ten years, twenty years from now somebody is going to come along and look at those drill logs and if you have made interpretations in recording the data, they don't know that. A new technology may come along that would make that muck into ore, if they know what the truth is. So, my contention is that you should record what you see and what you recover and let the engineers make the adjustments.

C. McLain

You are assuming that you recover all of the gold that is there, I assume it is all free gold or whatever your mineral is, it is all free mineral liberated and that you are taking everything right down to micron size.

J. Wojcik

Well, again, I would say make your very best guess of what is actually there and in that case, I'm talking about gravity recovery. If you are talking about alluvial deposits, you are making a judgment on the mining to end processing technique and even though you might use amalgam to clean up your sample, or go to a Gemini table or some such machine to make sure you are getting as much of the fine gold as you can. Again, document what it is, if possible do a size distribution on the gold that you recover, so the engineers know what they are dealing with.

C. McLain

Your figures were related to what is there and really have nothing to do with the mining method. You are just trying to see what is in the ground.

J. Wojcik

That's right, Chuck. Now, Lou has asked me if I wanted to address the various ways of handling samples, reducing samples. Actually there isn't much you can do. You classify them by sizing them and then you work over the finer fractions as you get into the smaller sizes. Ultimately, you are going to end up with either the gravity technique of panning or sluicing. You might then go to a dry technique or a wet magnetic separation. Some people like to avoid mercury and there are cases where mercury doesn't give you the right answer but you know you are limited only by your imagination and the high specific gravity of gold.

P. Mecklin, Independent drilling consultant

We had a job about 8 years ago in Indonesia. We were using cable tool rigs in a tidal swamp and what our problem was that we had so much water and so much beach sand that when we were using our cable tool rigs, we were getting theoretical volume or we were getting a volume that was coming up higher than what was the actual volume that we were penetrating. I was just wondering if there are any special tools for that type of situation that could be used.

J. Wojcik

You are not the first to ask that, Percy, and I am not going to answer it because rising sand is a terrible thing and the famous solution is just to drive through it. Generally you like to think there is very little gold in that nice clean sand with the hydrostatic head on it.

C. McLain

Were you leaving a plug of any sort in the drill casing?

J. Wojcik

Well, I have just given Percy credit for knowing what he is doing and obviously sometimes even leaving a plug you run into trouble.

C. McLain

Are you going to leave 5 feet of plug or are you going to leave 2 feet?

J. Wojcik

I just had a case in Wyoming a couple of years ago where if you left a plug, it drove ahead when you drove the casing and if you didn't leave a plug, you drove the casing ahead when you tried to drill the plug out.

C. McLain

I assume you were out of the rising sand when you left the correct plug. What about your method of sample interpretation?

J. Wojcik

In terms of sampling interpretation, I see a lot of cases where people will drill a fence of holes across a long sinuous supposed deposit and then they select all the holes that had gold in them above some cutoff and they lay that off and they start calculating ore reserves. You walk in and say let me see the cross sections. "What cross sections?" So how do you know what you are dealing with if you don't have cross sections? If you are going to drill fences of holes across a drainage, you have to look at your interpretation in cross section because sometimes you have to calculate your reserves in terms of the lithologic unit that has gold even though you may have holes that don't have gold above the cutoff. You are going to have to include those as part of your mining unit, part of your dredge course and you have to have a little bit of a geologic bent when you start calculating ore reserves from drill holes.

G. Saco, Canasta Resources

What both Joe Wojcik and Jim Noble say goes hand in hand. The size of plant, mining technique, the process recovery equipment starts in the exploration stages. My geologists normally hate me because I insist on a size distribution of the gold that is recovered from our sampling programs. Because immediately I start thinking about what kind of recovery system I am going to be looking at. As we go more into the engineering of it, we go into a pilot program and at this time you are looking at the characteristics, size distribution, and depth of gravels. You need this information to design the proper plant. I have been associated with the small plants to large plants from the arctic to the tropics, 50 yards an hour to over 400 yards an hour. You have to have the proper combination and it starts in the exploration stage. A classic example is the Livengood Bench, Livengood, Alaska. The cutoff grade on this property was 0.04 oz/yd^3 gold. We stripped 2.7 million cubic yards of frozen overburden in a period of three months and processed a half million yards of gravel in 2½ months. Recovered gold was a little over 14,000 ounces. At a cutoff grade of .04, you would expect to see at least 20,000 ounces of gold recovered in that season. What happened there, great earth moving equipment, great initial design on all the processing equipment up to the recovery system! Then sluice boxes with 20,000 projected, 14,000 ounces recovered, the sluices were only recovering at best 70 percent. Something went wrong, they went too cheap. Great equipment, great earth moving equipment, lousy recovery system! They lost the property through bankruptcy. The next set of lease holders came in and laughed all the way to the bank after redoing the tails. It takes a combination with the engineers, geologist and process people. You have to work together, you have to size the equipment properly, you have to know what you are doing. A good feasibility study is vital. Too many projects I've been associated with, the feasibility work is not done. They get their numbers, and think money in the bank. They go in with whatever they have, and first thing you know, there's nothing recovered Many good properties have been ruined this way.

L. Cope

Thank you, Guy. In addition to knowing the size of gold, I like to know the amount of heavy or "black" sands so I can intelligently design the cleaning stage of processing.

J. Wojcik

I would also like to reinforce what Guy says. That is, you have a preconceived notion about how you are going to mine the material and how you are going to process it, when you record your sample results. Later, someone may come along with a different process idea or different technique and be a lot happier with it and if they don't have the raw data to work with, they can't make the decisions. The other point is that when I was really vigorously consulting in the placer business, about the third week in August, my phone used to start to ring and the people would say would you come out and look at my plant and see what is wrong with it. We haven't recovered any gold and we have been mining all summer. My first question is, how do you know there is any gold there in the first place? What kind of sampling did you do before you started mining? They said well, everybody knows there is gold in the creek. The guys have been panning there for 50 years. And that is another point about placer sampling, you have to do the sampling before you start mining.

L. Cope

To follow along that same line, in placer, rural tropical areas, the local women will go down to the creek in the morning to wash clothes. While the clothes are drying on a bush, each woman has a little hole to dig at and they will do their panning. They just wade out into the creek with their wooden bateas and spin a little dirt. If they recover a quarter or half a gram of gold, they have done well and are happy. When the clothes are dry, they roll it up, put it in the batea, put it on their head and walk home. There is a case I saw on the east slope of the Andes where some people were working some small pits with wheelbarrows and sluices. A group of investors went in without any sampling, because the local people had been getting gold in the river. The investors set-up a considerable amount of equipment, but the gold wasn't there in that scale. There were just little hot spots the people were working in both examples

J. Wojcik

To cut in on Jim (Noble's) territory, one of the most clever and most unique small placer operations I saw was on a beach in New Zealand. This fellow had claims right along the beach and there was a little seam of garnet sand about ½-inch thick. The fellow had a wheelbarrow and a 55 gallon barrel cut in half length ways, a 12 volt battery and two 12 volt bilge pumps that he had punched into holes and styrofoam. He would load this gear in a wheelbarrow and go down the beach and find a little cusp in the beach where the sand was exposed and he would expose that little streak of garnet and then he would set up his little sluice box. He would shovel a wheelbarrow full of material and then wheel it up a plank to his sluice box and he had water in the two barrels he was recirculating with these little bilge pumps. His daughter would sit there with a large soup spoon and feed that muck into the little sluice box. One weekend I saw him produce a little over 1 ounce.

G. Saco

I will make a few more comments for you Lou. I can't stress enough that you have to examine placers in the proper manner. One of the major things you have to do is don't take the other man's word for it. At this last project I have been associated with, there was quite a bit of discrepancy of what we actually recovered going through the plant and what the original drill holes indicated. I was trying to find out what was happening and going back and examining the original data, there were great voids. As Joe has stressed. You have to be precise, you have to be accurate in how you record this information so the next man that comes along knows where these numbers came from. Ten million dollars was expended on one property that had a recovered grade which was half the original indicated grade.

L. Cope

Guy, you and I both worked on the same property at different times in South America which had the same situation.

C. McLain

I'd like to make a comment on the digging method. If you are going to use the type of digging mechanism that may classify before you get to the processing plant, I think you did make mention that you can lose values on the bottom. Where a mineral is easy to recover in your recovery system, it is going to be just as easy to loose on the bottom.

L. Cope

Thank you Doug (Piper), you have given gave us some excellent information. In other words, the other follow's business always looks easy, that is, it doesn't just look easy, it IS easy. But the key is to anticipate the problems before they hit you and that is what Doug has given us. You did dwell on that item of lack of planning. I think you are the third speaker in a row to do that.

J. Wojcik

A question that was never answered at the Perth Alluvial Conference was, "What sort of guidelines do you use to determine what sort of cutter horsepower is needed?"

D. Piper

That's a good question. I think the real answer is that it is as much art as science, I am not sure we have a formula we could put on the board or anything. The main problem that we have is getting any kind of information to make any kind of an informed decision or informed guess about. When I heard Joe talking about the pit sampling versus the core boring of whatever, I had to make a note here because one of the things we like to see is somebody dig a pit and we like to have all of the details of the equipment they use to dig the pit and what the production rate was when that particular equipment dug the pit and whether it had to strain and grunt and groan to dig it or it went easily and how things worked. These bucket wheels, especially the small bucket wheels, have a hydraulic drive and we can calculate digging pressure or tooth pressures on that rotating wheel and if you dig a pit, for example, with a back hoe and can get similar type of data from the backhoe, then obviously similar tooth pressures, similar lip pressures on our excavating wheel are going to give similar results. From a statistical sense, the pit may not solve a lot of your problems. From our viewpoint of trying to get a handle on what it takes to excavate, it solves an awful lot of ours. The biggest problem we have with people coming in to buy a dredge and, in fact, you know we probably get an inquiry like this once a month if not once a week, is "I want a dredge". What size? " Well, I don't know I want to dredge some stuff." We need some data especially on the geotechnical aspects of how hard is it, how cohesive is it, what is the sheer strength, did you do any SPT when you drove for your cores, how do we make a guess on what it takes to excavate this. I had a customer come in and said they had 2,000 holes for the determination of all the mineral properties, where it was, the limits of the ore body, how it was going to be recovered, what size plant they would need. They had a deposit good for 20 year mine life and they wanted a dredge that was going to be in the 8-10 million dollar range. Only six of the 2,000 drilled holes had any geotechnical data to decide on how much horsepower you needed to dig it. **That** is a problem.

The bucket wheel traps the material better and does clean the bottom better and in hard material better than a cutter head. Those of you familiar with dredging at all will know that rotating excavators dig very well in one direction and not very well in the other direction. The bucket wheel digging from the bottom digs equally well in both directions and, therefore, in harder material you get a more uniform and higher production rate. It is more expensive, it has a little more maintenance than the old rotary cutter but it performs better in picking up the heavy minerals, it handles clay better in most cases. You can get a higher through put in clay with less clogging.

P. Mecklin

I was on a job about four years ago, and like I said at the time, it was a drilling problem. These people were off the coast of New Zealand and they were wanting to sample a project that was under about 300 feet of water and about 8 miles off shore. There were high waves and high tides. My comment to them at the time was that there isn't a drill available but we can design one. We offered them the mechanics of making this drill but, I told them at the time. I said, my concern if I were you is how you are going to mine it even if the values are there. It ran high grade, probably about an ounce per yard or something like that so it looked good They said at the time they weren't interested in putting their money into it unless they could get a joint venture. I was just wondering if there is any type of equipment available or what would you suggest on something like this?

D. Piper

If we could get them to put their money into it, we would work on some equipment. No, there have been studies done. I'm sure you have all read and probably have done as much about it as I do, some of this deep ocean nodule mining and things like that. We have looked at something similar, generally in 100-250 feet of water basically by using clamshell-type apparatus and a hopper dredge. There are some schemes around for doing things like that. I don't think any of them have been built and you would want to bring plenty of money.

P. Mecklin

Is something like that feasible?

D. Piper

I think it is feasible, but again you are talking about investments 25-50 million dollars and you haven't got the muck on shore yet. You might be able to screen it out there if you had a big plant and do something like the placer miners do with bucket ladder dredge, trommel, screens, and jigs.

I'm sure you would be able to do all that floating offshore. Again, if your size and money was right, but you know the first cost would be certainly in the 25-50 millon dollar range for the piece of equipment, something like that.

J. Wojcik

The other point that people don't seem to recognize when they want to talk about deep water, is when you have outlined a mineral deposit that you want to mine, you want to get the maximum extraction of that deposit that you can. When you are in 300 feet of water, how do you position the barge. If you are held in place with anchors, there is slack in your anchor lines. The barge is going to move. If you use a clam shell, are your going to go out and dig a bunch of bomb craters and come up with 15-20 percent extraction? That is a serious problem as those things have to have very precise positioning. One of the reasons the bucket wheel has been effective and has been successful, particularly the Ellicott, is because of the movable spud carriage where you can have a very precise incremental advance of your cutter into the bank.

G. Saco

Doug, on a little smaller scale, these bucket wheel attachments that you have for backhoes, I think you have three sizes of them. Do you have any price information with you on those of your standard sizes at this time?

D. Piper

Right now we are selling those as sort of a package. You need a hydraulic power unit and the excavator itself with the pump built into the back of it. So these things are really a combination excavator and pump where you can pump a few hundred feet let's say to some kind of a screening apparatus or just holding area or depending on what you are going to use it for. But the whole package is probably in the order of magnitude of a quarter of a million dollars.

G. Saco

I have some literature from you that was a retrofit for backhoes. It was just a cutter head.

D. Piper

What you think is a cutter head has a pump built into the back of it.

G. Saco

Yes, it has the hydraulic pump built it into the back of it. Right, and that is supplied off the hydraulics of the excavator.

D. Piper

No sir, no.

G. Saco

It isn't?

D. Piper

No, you have to buy a whole separate unit consisting of a cutter head and a hydraulic pump. These things require probably 200-300 horsepower to operate it depending on the size of the bucket wheel. We have them running. The excavator alone in the say 50-100 horsepower and the pump to pump that type of material away probably has got to have the same or a little more horsepower so you end up needing 250-300 horsepower total to dig and then pump it.

G. Saco

Can't use the hydraulics pump off the excavator?

D. Piper

No, you can't.

J. Murray, Arrakis Corp.

I noticed you quoting capacity on these dredges and I have been tripped up on this before so I wonder if anyone else has here. 2,200 tons per hour of what?

D. Piper

Bank material.

J. Murray

That is solids?

D. Piper

As solid as the bank is.

J. Murray

Ok, so you are not taking into account a slurry density then. What do people normally see in terms of a slurry density in the line going out the discharge?

D. Piper

Again, that can be all over the place depending on your equipment and your project, but the big dredge I showed here with a wheel excavator digging a high bank of fine sand and a submerged ladder pump is running slurry densities pretty continuously between 1.3 and 1.35 specific gravity of the slurry.

J. Murray

Something which I've found quite often when people quote machinery rates and obviously you take this into account. When you start talking about how hard it is to dig is also a function of what kind of slurry density you are going to maintain because it is going to pump "x" thousand gallons of water no matter what you're feeding it. So people who don't take that into consideration all of a sudden, for $8 million they've now bought a machine that only gives them 500 tons per hour instead of a thousand tons per hour. It has a major impact on the cost per yard and the economics. And I guess more of a comment, have you found that to be the case?

D. Piper

Absolutely. You get a couple problems, but as you said, the real secret is to pump it as slowly and thickly as you can, as far as your energy and wear are concerned. I'd try to keep the velocity in the pipeline down, and the specific gravity up, and that's the whole secret of the business. My paper has thrown around some pretty high wear rates in gravel. That happens when you aren't pumping any solids, and the solids you are pumping are cobbles and are eating away the cutter and the pump faster than you can replace them practically. Then obviously, you're production is down and wear is up. You'd want to do just the opposite, to dig it as thickly as you can and pump it as slowly as you can. And, again, we just had some people in our office that are contractors who have a lot of cutter dredges and a couple of wheel dredges. We asked them what their experience has been. The comment was that they could get about twenty percent higher density in slurry by using a wheel. That's sort of their rule of thumb. And that converts right into energy efficiency and wear. But the other comment in this paper about everybody choosing equipment that is right against the limit. Alex McDowell has been very successful selling equipment for successful projects simply by taking the nominal rating of the equipment and dividing it in two or some factor like that. Really, if you're going to be operating for a 20-year life, you're going to have all kinds of problems and you can guess that probably all the geotechnical work and everything hasn't been done, you're a lot safer to buy a unit a little bit bigger than you really need. Then, when you get there, your problems are not with the dredge, they're with the plant or something else.

T. Ellis, Ellis International Services

Back to the ocean mining problem. Just of interest, there is a company in Australia that recently reported finding a very large diamond field in the ocean off the northwest of Australia, Kimberly District, where the existing diamond mines in Australia are at. They are hoping eventually to get production something in the order or about 15 percent of the world's diamonds. That might be a really good one to chase down for somebody interested in development of underwater mining.

D. Piper

Thank you, Trevor. I can also comment along that line to get you a feel for these depths. We are actively pursuing a project now where the dredging depth would be 45 meters, about 150 feet, and that is considered pretty deep. We have designed some equipment. We were not the low bidder, it wasn't built, the project died but we had designed some equipment to go down to 60 meters. But after that, you really start looking at equipment that needs to be a little innovative and really hasn't been built before. I would say at about 45 meters you can come up with reasonably standard equipment.

R. Dalton, MSQ Co.

You have been talking about the geotechnical data that you need for dredging. Why don't we put it on the record and see what you need. Tell us what geotechnical data you need to help you plan the dredging.

D. Piper

Well, we need as much as we can get. We are accustomed to working with blow counts from SPT, sheer tests and compressive strength tests if the material is in fact indurated or cemented material. Again, if we get into something like this where we have any doubt, we usually try to prevail on the customer to dig a test pit if that is at all possible. At a project we just completed, the customer put a drag line in for a couple of weeks both as a sampling and as a means of checking the difficulty of excavating various layers. Again, if my associate, Alex McDowell, were here, he would tell you, "dig it with a backhoe and I'll tell you what you need to dig it with a bucket wheel." This is primarily based on tooth pressures and lip pressures.

J. Wojcik

I want to comment that in processing, particularly, is where we see a lot of cart-before-the horse sort of thing. Again, my phone rings very often. A fellow says, " I've invented this machine to recover gold. Do you know some place I can take it to put it to work?" Recovering gold is fairly simple if you stick to the basics. There are a lot of tinkerers out there that just love to invent things. But you have to have the deposit first before you can process it.

H. Hendricks

I am interested to know if any members of the panel could tell me what the proper process for sampling of a terrace placer is today, in 1991.

J. Wojcik

Well I guess I would have to reply that I would have to see the specifics of the deposit because terrace placer takes in a multitude of sins and in that case in deference to Chuck McLain's comment earlier, you do have to think a little bit about what you're mining method is going to be and what your mining costs are going to be. Certainly you are not going to put a floating dredge up on a terrace gravel and so you are going to be thinking about higher cutoff grades and different distributions. I would have to see the specifics before I could say a whole lot.

H. Hendricks

I would like to ask considering terrace gravel deposits, in the event that you had a, let's say a trommel, a grizzly, a small pilot plant capable of 25-30 cubic yards bankrun material per hour throughput, would it be recommended to run small bulk samples as opposed to drilling, or do you feel that drilling would give you better data?

J. Wojcik

Again, you have to back up and get in your mind's eye a picture of what do you think was the mode of deposition of this deposit? Is it somewhat of a lag that migrated across the valley at one time or is it likely to be channel-like. And those would determine whether you wanted to use multiple points of smaller samples or fewer points of large samples.

H. Hendricks

Let's say that it's channel-like.

J. Wojcik

If it's channel-like, first you have to locate the channel and if you are confident that the channel is continuous and uniform and it can be reached with pits, pits are certainly preferable to drill holes in that you can actually see the material, you can get an idea of what excavating horsepower is going to be required, but obviously you have to do some drilling before you can figure out where the channel is and whether it really is in fact continuous.

H. Hendricks

What if you have a channel which has been opened by previous underground workings and you can follow and see the channel in place in the underground workings? Would you still recommend a drilling program or the bulk sampling?

J. Wojcik

Well, then you have the additional problem of how much of it is gone and that is a big problem.

R. Mundell, U.S. Bureau of Mines, Denver

I was just wondering, I got involved with some black sand deposits a number of years ago on my own. It was material with good values, free gold, but I was having to contend with all 200, 300 and 400 mesh gold. And, I am just wondering here we are, 1991, and we have a lot of experts here, but where are we with the state of the art to handle this fine of a material with free gold of 300-400 mesh? What would be your first reaction to that?

L. Cope

There are machines, I feel that it depends on the type of muck and how slow you want to run. Just like running a trommel. If you have a lot of tacky, sticky clay, you have to run a lesser amount of material. The centrifugal machines and spirals will collect this fine gold. But you have to screen to feed them, and you have to cut the capacity.

On the Snake River in Idaho, the gold is known for its very fine, flat, platy particles. There was a gravel plant that recovered the fine gold. Now this was a quite favorable situation because they did all the screening and washing and then there were spirals for the gold recovery. Another example is, at the mouth of the Carate River on the Osa Peninsula in Costa

Rica with the beach mined between high and low tide. Jigs were recovering very fine gold. I could look at the gold in the hutch concentrate and it would fade out smaller and smaller and until you couldn't quite see it. You could put your 10 power lens and see that it got still finer until it was too fine to see, again. Now I don't know how fine that was, but I feel the reason that this was recovered was that a natural bed developed in the jig. There was a lot of fine gravel which was coated with manganese and iron oxides. This slightly heavier material in the jig bed is what I attribute to recovery of the fine gold.

J. Wojcik

How about percentage and efficiency?

L. Cope

I have had a lot of problem with talking about percentage gold recovery from an operation, because an operating plant is so very difficult to sample. Besides, a little bit of a nugget effect will go a long way to upset the results. In addition, I feel that in many cases, excessive effort to go after recovery of ultrafine gold particles is self-defeating. How many "jillion" of these ultrafine particles does it take to equate to any dollar value?

B. Knelson, Knelson Concentrator Co.

I want to point out that I can recover 400 mesh gold in my sluice box whenever I shut the pump off, but what percentage of the 400 mesh gold that goes through the box did I really recover?

I have a couple remarks about the hydrophobicity of gold below 50 microns. You are absolutely correct when you are talking one gravity, but that curve changes with multiple-g's. There is a report by McGill University that gold down to ten microns is readily caught by a Knelson Concentrator, though I can't give you a recovery figure.

I also wanted to comment on some of your remarks, Lou. I can catch ten micron gold in a gunny sack. But the question is how much of it? The thing I will talk about for a second in sampling, and I find it so common in the industry, is that you get all these "whizbang" machines which people have created, including mine, where you go onsite, or the people bring a sample in. If you want to plan a bit of the "Carnival Caper," you run that one little sample through, and the people say, "marvelous, great recovery." If you're going to sample in the field, if you're going to spend some big bucks on what you're going to do, make absolutely sure that you're system is clean. I was just talking to two fellows who had gold in their first samples that didn't come from their site. It came out of the machine, not our machine, but another one. The second thing is, run it for a decent length of time. Don't go with little short samples and think you are going to do that and/or repeat it in the field, because what will happen is that in the system, whatever it is, will load and will start to kick the gold through. Run something in a comparative manner to how you're going to run when you're in operations. Don't go sample with a Knelson Concentrator, then go with a sluice box, because it "ain't" going to give the same results. They're two different methods--like apples and oranges.

Clay problems are something, there's more and more technology coming out on clay. And clay will beat a lot of systems. And so you have to be sure you run those a decent length of time with sufficient volume to give you a good feeling as to what you're really going to recover when you do go into operations.

L. Cope

Thank you for those comments. That's what we really want to hear, the "how to" remarks.

H. Hendricks

I'd like to put in a plug for Jim Vickery of Aerosort, which has a technology that Mr. Vickery showed here at the exhibit, with regards to its ability to capture microfine gold. I have personally run quite a number of samples through the machine, and we have been able to recover particles of gold to minus 1500 mesh, dry processing. When you get into the micron particles sizes, the minus 300 and on down, you have a floating factor with wet systems which you don't have in the dry system. Its the buoyancy factor.

G. Saco

As Lou said, a major factor in the size of gold recovery you're going after is the economics. You can get gold down to a very small size, but it usually takes a loss of capacity throughput to get it. Therefore you're operating costs are going up, and its as simple as that.

L. Cope

An excellent point by a well experienced operator.

C. McLain

I'd like to make a comment on practical recoveries. Doug, you were faced with the problem of "what are you losing off the back end of the dredge?" In one situation I was involved with, we tabled bulk tails samples and panned that concentrate, and called it the recoverable concentrate. Then we amalgamated everything else and called that what we would like to get, but we don't know what the size was. We acid-leached the amalgam and tried to estimate from the particles remaining what the size might have been, and it was just a guess. Finally there was cyanidation. We would cyanide a batch to see if we could get the last squeak out of it, and find out if there was any non-amalgamating gold. We didn't get that far, but you can see that we weren't really getting a size analysis on that gold. We were getting a recoverability analysis, and we still didn't have a size analysis. This is a challenge for anyone who is looking for challenges. What are the losses in these plants?

J. Wojcik

Perhaps Mr. Mundell from the Bureau of Mines would like to comment on some of the Snake River gold on the Fort Hall Indian Reservation. There's still quite a bit of contention about that gold. Apparently it's a naturally occurring amalgam in a very fine particle size. It doesn't amalgamate in a pan, and the only way they've been able to recover it from their samples is with flotation. Now there's something to think about.

S. Dix, Bond Gold

I've worked on some Tertiary gravels, and I would like to know what you think of our sampling method. We first went in with a 9-inch Becker Drill and the sample reduction in a Gold Saver with a trommel. We then ran the sluice tailings into a 6-inch Knelson. We screened at 28 mesh and hand picked all the plus 28-mesh and panned all the minus 28-mesh on the sluice. We floated the Knelson Concentrate and then panned that to give us a general idea of the distribution we had on it. We basically felt that anything which wasn't floatable wasn't really worthwhile. We did the whole thing without any mercury in the sample at all. We then went in with a 24-inch clamshell and reproduced the grade of the gravel itself. Do you have any comments on going to that procedure?

L. Cope

Actual froth flotation?

S. Dix

Yes, with amyl xanthate.

J. Wojcik

My initial comment would be that this is exploration. It was a reconnaissance program basically, to establish the existence of gold, or not, in significant quantities. It is questionable if you could afford to follow that procedure on a scale that would be necessary to outline a minable ore deposit. The economics might preclude you from going to that trouble. But it certainly is an excellent way to find out what's there, and if you document the procedures which you used, then engineers ten or twenty years from now could look at your data and understand what you did.

L. Cope

I'd like to add that I like to see values recorded not in cents and dollars, but in milligrams per unit volume. Sometimes you don't know what the price of gold was when the sampling was done for those old reports.

L. Nonini, U. S. Bureau of Mines

I didn't work on the Fort Hall project, but did have experience on some of the Snake River material, with Professor William Westley Staley about 60 years ago. At the time we were working with that sample, about one flake out of ten would be white. Staley claimed that was platinum. I don't know what he based his conclusion on. I never saw an assay or any definite information other than Staley's work. But I was thunderstruck to find out that a couple years ago that this was a natural occurring amalgam. You seem to have some interesting thoughts on that. I just wanted to mention what Staley thought it was.

L. Cope

We appreciate your bit of history, Mr. Nonini.

J. Wojcik

To follow-up on Lou's comment on recording sampling result units. There was a case here in Colorado of some people who spent a significant amount of money on a deposit where the values were reported in "gr. gold." They assumed that these were "grams." After they got on the property and spent some money, they found the old report gave gold value in "grains." So again, record your procedures carefully.

L. Cope

Amen.

J. Wojcik

Last Fall Dick Tinsley in project finance with the Indosuez Bank of Sydney, Australia, gave a paper on why alluvials are not bankable. And one of the points he made was that it is so difficult to show proven and probable ore reserves. And further on comments about placer deposits in the Soviet Union by Mort (Richardson), I just returned and one of the reasons we don't have proven and probable ore reserves in placers very often is because it is just too expensive to drill them and sample them to that point. With the low grades that you encounter, it is just not economically successful. In the Soviet Union, however, economics don't enter into their gold mining plans at this point. They receive a budget and they have a quota and the quota is not related to the budget directly. So they do drill out their reserves on a very methodical manner and it is very interesting.

R. Dittrich, Metallurgical Consultant

I had the opportunity a few years back to do quite a bit of work in gold placer mining both in the Western U.S. and in the Kilometer 88 region of Venezuela. A couple of questions come up. Sluices might in fact be out of style, but I used one in Arizona that got 90 percent recovery on the entire spectrum of gold size. I did some detailed checking on overall gold content and we were still getting about 50 percent of our overall minus 400 mesh gold on a sluice. A question for all of you. I have heard comments about amalgamators and I will give you a question for which I found some information but not enough. When I was in South America, it was a common practice to throw mercury into the hole as they were getting down to their gold bearing gravels and then they would run it across their sluice, and then do their clean-up. My question is, do any of the gentlemen on the panel have information regarding the solubility of gold in mercury considering the fact that some of these fellows would put 2 kilograms of mercury into the hole and only come up with about 300 grams at the end? I calculated a loss of gold due to solubility of over 50 percent. Comments?

M. Richardson

That again, is part of this irresponsible mining procedure being carried on by the garampiero-type miners throughout the world. These people are killing themselves and others with this loose handling of mercury. We certainly wouldn't endorse any aspect of it and of course Chuck McLain could point out that the dredge he used to manage, the Yuba Dredge, was a net producer of not only gold but also mercury left over from the early days going back to the early part of the century when riffle boats were used where the procedure was to spread mercury over the sluice itself. You know how much of that mercury had to go into the water.

R. Dittrich

One other quick anecdote. In a property in the Mohave, we found those nice platinum-appearing specks of gold in with our gold and upon further investigation we found that it had been mined in the 1920's and a lot of people shot at each other. It turned out that the white shinny particles were lead!

J. Ranserhoff, Neutron Products

We are producing gel-logs which have been used in placer mining in Alaska. These logs can do an effective job in water clarification in the 20 to several hundred parts per billion range. In your fearless forecast of the future, do you see a greater use of flocculants in clarifying the effluent from placers?

M. Richardson

One of the problems of adding anything to the water when you are dealing in large volumes like I was talking about, it really becomes impractical. Maybe you are working either in a large flowing river or you are working into the benches and the river flats that carries the river behind you. There is so much water circulating that it isn't practical. Typically, these big producing dredges were dredging in muddy water and still had very high recovery and it wasn't economic to fool around with this very, very fine gold but the majority of the gold recovered was what most people would consider to be fine, not nuggets.

B. Knelson

A couple of remarks on the things you are talking about. You mentioned the Bulolo River. I am not quite sure which property it was but one of the problems that you run into in attempting to prove what can or can't be done is purely political. We had a situation where the natives were mining below the operation on the Bulolo. They were making a living on tailings. We went to that particular mine and said look, can we run some of your tailing and show you what is happening. The foreman said get the hell out of here, we don't want to know what we're losing. As long as we don't know, we don't have to hurt, so just get out of here. He wouldn't let us touch it. The second thing is that I heard you remark that with the invention of jigs, riffles should no longer be in existence. I do want to tell you that we have replaced 17 jigs in the Northern Territories of Australia in alluvials. The people running them have informed us that in the dry season with jigs as the water got dirtier, their recoveries went down. When they put in the Knelson concentrators, the recovery stayed flat, the increase in recovery was something like 35 percent over the jigs and we are replacing jigs fairly consistently in hardrock circuits at some very well known names. We are talking about Montana tunnels and some of Noranda's mills, Campbell's mills, *etc*. I guess we are in about 8 to 10 of them now and in probably 80 percent of them we have replaced jigs. Some of them were ones that didn't really have a gravity circuit until they tried our units and found that they worked. I just thought that you should be aware of that because it is taking place.

M. Richardson

Yes, we're aware of that, and we're also aware of the fact that there are very few people around these days who understand how to tune and operate a jig properly, so that is also part of the problem.

J. Ranserhoff

I don't think you understood my question. What I am asking is, in the operation of placer systems, do you see any need in the future in this forecast for clarifying the effluent?

M. Richardson

I understood what you meant about flocculation. Yes, there is no question there is a lot of testing that has been done with that. I think in smaller scale mining, it very well could be appropriate. But if you are in higher volumes, bucket ladder dredge-type operations with 3-500,000 cubic yards a month, generally there is just too much water going through that dredge plant to make a flocculant necessary or effective. But you know I think there is always a possibility of things like that.

J.Ranserhoff

I am telling you that with the log and the stream method, you can get effective flocculation in the 20 to 100 part per billion range.

M. Richardson

Yes, but what if you are circulating 30 to 40 thousand gallons of water per minute through that system?

J. Ranserhoff

You don't have any overflocculation, so that the utilization of floc is very good, and you're literally getting decent results in the dozens of parts per billion range. And I think its economically feasible to get the clarification. The question is, is there a demand for it in your foreseeable future?

M. Richardson

Again, a system which was developed back in the 1930's with the dredges in California, was to put a suction hose down just behind the ladder. Of course you understand that the best values to be found are in digging in the lower levels. So you might be digging anywhere from 50 to 100 feet below the water line before you are really getting into the values. Now, by putting a suction hose down there and spraying that back over your tailings, you're taking the slimes which tend to build-up in those lower regions, and you're clarifying the water. Now that has been the most effective way to handle this kind of a problem. To try and add some kind of a colloidal additive in water in that volume and getting it down to 50 and 100 feet and into the area of all this digging and agitation which produces a tremendous amount of turbulence, I believe, is a little bit pushing the system. But if you get into something like a backhoe operation feeding a floating washing plant of maybe 50,000 cubic yards a month from a confined pond, there could very well be a good application, and there will be more and more of that type of operations.

S. Dix

Going back to the fine gold in the Snake River, I saw an article a few years back of using the air-sparged cyclone on some of the Colorado fine gold. This machine was developed to process large volumes of solution with just a small piece of equipment with very short residence time. I have talked to Chuck McLain about using it on the Yuba dredge. I don't know if you were able to get around to trying it Chuck. We're trying it at Coliseum for floating off sulfide minerals. Very little floor space is required with it. An example they gave was a 6-inch cyclone was comparable to a 150 cubic foot flotation cell. Has anyone even known about the piece of equipment and looked at it for the fine gold recovery of setting up rough or cleaner circuits.

C.McLain

No I haven't heard of it.

B. Knelson

I am not that familiar with some of this processing equipment. The University of Alberta spent a great amount of money on the compound water cyclones. Is that the same thing that we are talking about? If it is different, then because they spent about 3 million dollars to wind up with a zero, I don't know if that would have the same problem or not. The one thing that I can tell you is the difference between the centrifugal concentrator and a cyclone. Everyone thinks they are the same, because they both go round. It "ain't" the same. They are absolutely diametrically opposed. A cyclone separates primarily by particle shape and size and secondarily by specific gravity. A centrifuge separates primarily by specific gravity, secondarily by particle shape and size.

S. Dix

Just to get the air-sparged cyclone clarified, it uses cyclone force but the cylinder has porous material on it where air is sparged into it and your overflow is a foam coming off, so you are making a flotation concentrate and it is not really going in there being separated by specific gravity and/or particle size. You are separating it by the froth flotation action itself and I think it would be something to look at for the fine gold because with the rougher cleaner circuits, you could probably reduce the volume considerably. I think, Chuck, when I talked to you one time, the fine stuff was real easy to amalgamate, if I remember right. I know it takes a lot of fine gold to make any money but it is a piece of equipment that doesn't take much floor space at all and it is just pumping costs on top of that and you are talking very low pressures for blowing the foam off.

L. Cope

I have a question; any reagents at all used?

S. Dix

I would think you would probably want to look at that but with the natural hydrophobicity of it, you may be able to get away without anything. But I have only seen one paper on it and I think it might be something for more investigation to see if you can get away without it at all.

M. Richardson

Kind of adding on to that idea, one area that we are trying to get across to the spiral people is the suggestion in all types of heavy minerals is that they really ought to look strongly at a primary jig preceding their spirals. One of the difficulties we are seeing for instance where they even try with gold or other minerals is that the volumetric throughput in even a huge nest of spirals is really not adequate to keep up with a large scale dredging operation. The rougher jig can perform the function getting the heavy materials concentrated, get rid of a lot of the waste materials and then let the spirals do their job. I think it would improve efficiency tremendously.

L. Cope

I want to amplify on that, Mort. In addition, by using a jig to preconcentrate before a spiral, you also get the screening effect.

M. Richardson

That's right. You can put half-inch material over a primary jig, but for a spiral you have to screen it down to $^1/_8$ inch. Also, you have to worry about organic matter and things like this which the jig is ideally capable of removing. I think it would be a great improvement to spiral plant systems for all types of minerals.

L. Cope

Pictures come to mind of the spiral barges with literally thousands of spiral turns. These are several stories high, and are actually mineral sands plants on enormous barges. These are still being used for mineral sands recovery.

R. Dittrich

One comment regarding the essentially "flotation" hydrocyclone. That paper was given at one of the sessions here in the AIME Convention. You might check the preprints for further information on that device. Going to a whole different subject, that of water quality, water clean-up, use of flocculants, *etc.* The discussion dealt with the barges and the floating-type processing plants. In Alaska, particularly in the Denali Wilderness a few years ago, there was significant work done both on grant and independently by private companies that was published and is available. It was done on a ponding filtration-type system. That system and that information were used very effectively to meet all water quality requirements for the state of Alaska and I think there is a gentleman who might give you further directions on where to write for that. I don't have them with me but, but I happened to be involved with the company a few years ago that had some of that work done privately and I know that other work has been done by The University of Alaska.

C. McLain

If the general public thinks that we don't pay any attention to what we discharge, that is certainly a mistake. In some nations your discharge is very closely regulated. We have a lot of experience in flocculation.

Every slime is different and every slime takes a lot of research and you try to treat as little as you possibly can. But if you are in the middle of your process, you can be treating hundreds of thousands of gallons a minute. That's a lot of water and you generally are not adding flocculants! I have a question directed to Mort. And that is, you are looking to the future, what do you think of getting deeper? Everybody is looking at 60 meters, maybe as deep as 100 meters, do you think Norman Cleaveland's endless drag line is going to get us deeper?

M. Richardson

I'm convinced that the continuous drag line that Norman Cleaveland, Pat O'Neal and Roland Gunnert worked on for so long, needs to be tested. However, I am pretty persuaded by the arguments that Romy Romanowitz made against it. So I question whether that would be the proper tool, but it might be. Aside from that, the only system, and the system is available today, is like they have used in Holland and a few other countries where they are going very deep for sands over 300 feet with just a dangling suction pipe and a submerged pump down at the end of it. Now there you are almost unlimited how far you can go but, of course, you are never going to be able to clean bedrock. If that isn't a problem, if there is just a huge alluvium deposit that has, whether it is cassiterite or gold or some other heavy mineral, something along those lines, is going to be the way to do it. But that technology is here. Going down much deeper than that though, you know then you really start to get a lot of questions and of course if you are too far off shore, if you are thousands of miles, like at the manganese nodule fields, the metallurgy alone I think just kills it. But that transportation logistics is also a major cost. I don't know, it just depends on the deposit. As I say there are still a lot of deposits probably at 150 feet that could be mined before you start worrying about the deeper ones.

L. Cope

Well, I think we ought to wrap this up. It has been a long four days of conventioning, partying and learning. My appreciation to the other panel members, and I appreciate the audience participation.

SELECTED BIBLIOGRAPHY

Anon., 1990, "Pontoon Dredge Design Uses Inventive Jet Pump System," *Rock Products*, Vol. 93, p. 17, Jul.

Anon., 1988, "Pilot Dredge Proves Gold in Ecuador," *Engineering and Mining Journal*, Vol. 189, p. 48a, Dec.

Anon., 1983, "Methods and Costs for the Attainment of Placer Mining Guidelines Effluent Limits," Environmental Protection Service Yukon Branch, Environment Canada, 42 pp.

Anon., 1979, "Conference on Alaskan Placer Mining," Mineral Industry Research Laboratory Report No. 43, School of Mineral Industry, University of Alaska, Fairbanks, AK, 138 pp.

Anon., 1952, "Yuba Dredges," Yuba Consolidated Industries, Yuba Manufacturing Co., Benicia, CA.

Anon., 1946, "Yuba, World Leader, Placer Mining Dredges," Yuba Manufacturing Co., San Francisco, CA.

Anon., 1928, "Tisco Manganese Steel Castings as Applied to Gold, Platinum and Tin Dredges for Alluvial Mining," Taylor-Wharton Iron and Steel Co., High Bridge, NJ.

Anon., 1915, *Dredges and Gold Dredging*, New York Engineering Company, New York, 68 pp.

Anon., 1915, "Yuba Gold Dredges, Yuba Ball Tread Tractors, Yuba Irrigation Pumps," Yuba Construction Company, Marysville, CA.

Anon., 1912, "Placer Mining Dredges for Gold and Tin," Union Construction Company, San Francisco, CA.

Anon., 1908, *Placer Mining; Surface Arrangements at Ore Mines; Preliminary Operations; Ore Mining; Supporting Excavations; Assaying*, International Correspondence Schools, International Textbook Co., Scranton, PA.

Anon., 1903, *Gold Dredging Machinery*, 5th ed., "Catalogue No. 17," Risdon Iron Works, San Francisco, CA, 50 pp.

Anon., 1897, *Placer Mining*, Colliery Engineer Company, Scranton, PA, 146 pp.

Averill, C. V., 1946, "Placer Mining for Gold in California," California Division of Mines and Geology Bulletin 135, San Francisco, CA,

Boericke, W. F., 1936, *Prospecting and Operating Small Gold Placers*, 2nd ed., J. Wiley & Sons, New York, 144 pp.

Braithwaite, 1899, *Auriferous Otago: Our Dredges, Where They Are and What They Are Doing*, Dunedin, New Zealand, 47 pp.

Cook, D. J., 1983, "Placer Mining in Alaska," Mineral Industry Research Laboratory Report NO. 65, School of Mineral Industry, University of Alaska, Fairbanks, AK, 157 pp.

Debicki, R. L. (ed.), 1983, "Yukon Placer Mining Industry, 1978-1982," Exploration and Geological Services, Northern Affairs Program, Indian and Northern Affairs Canada,

Douglas, J., 1948, *Gold in Placer*, W. Hebberd, Santa Barbara, CA, 123 pp.

Dutra, E., 1976, *History of Sidedraft Clamshell Dredging in California*, 2nd ed., Dutra Dredging Company, Rio Vista, CA.

Fansett, G. R., 1917, "Sampling and the Estimation of Gold in a Placer Deposit," University of Arizona Bulletin No. 51, Tucson, AZ, 17 pp.

Gardner, W. H., 192?, *Drilling for Placer Gold*, Keystone Driller Company, Beaver Falls, PA, 196 pp.

Gomes, J. M., et. al., 1979, "Recovering Byproduct Heavy Minerals From Sand and Gravel, Placer Gold, and Industrial Mineral Operations," Report of Investigations 8366, U.S. Bureau of Mines, 15 pp.

Graves, T. A., 1939, *The Examination of Placer Deposits*, R. R. Smith, New York, 168 pp.

Hanneman, K. L., et. al., 1987, "Placer Mining Demonstration Grant Project Design Handbook," L. A. Peterson & Associates, Fairbanks, AK, 46 pp.

Jenkins, O. P., 1970, "Geology of Placer Deposits," Special Publication No. 34, California Division of Mines and Geology, San Francisco, CA.

Johnston, S. P., 1927, *Gold Dredges*, Yuba Manufacturing Company, San Francisco, CA, 44 pp.

LeBarge, W. P., and Morison, S. R. (eds.), 1990, "Yukon Placer Mining and Exploration 1985-1988," Exploration and Geological Services Division, Mineral Resources Directorate, Northern Affairs Program, Yukon Region, Indian and Northern Affairs Canada, Whitehorse, Yukon

Macdonald, E. H., 1983, *Alluvial Mining, the Geology, Technology, and Economics of Placers*, Chapman and Hall, New York, 508 pp.

Macdonald, E. H., 1972, *Manual of Beach Mining Practice, Exploration and Evaluation*, 2nd ed., Australian Government Publishing Service, Canberra, 120 pp.

Maravi G., M. E., 1988, "Probabilistic Risk Analysis in Evaluation of Gold Placer Deposits in Puno-Peru," Master of Science Thesis, Colorado School of Mines, Golden, CO, 272 pp.

Masson, D. L., 1953, "Small Scale Placer Mining," Washington State Institute of Technology Bulletin 45M, Washington State University, Pullman, WA, 64 pp.

Mertie, Jr., J. B., 1975, "Platinum Deposits of the Goodnews Bay District, Alaska," Professional Paper 938, U.S. Geological Survey, 42 pp.

Minard, J. P., et. al., 1976, "Alluvial Ilmenite Placer Deposits, Central Virginia," Professional Paper 959-H, U.S. Geological Survey, 15 pp.

Orris, G. J., and Bliss, J. D., 1985, "Geologic and Grade-volume Data on 330 Gold Placer Deposits," Open-file Report 85-213, U.S. Geological Survey, 172 pp.

Pacheco, J. E., 1975, *El Principio del Placer*, 2nd ed., J. Mortiz, Mexico, 163 pp.

Prelini, C., 1912, *Dredges and Dredging*, 2nd ed., D. Van Nostrand, New York, 279 pp.

Reimnitz, E., and Plafker, G., 1976, "Marine Gold Placers Along the Gulf of Alaska Margin," Bulletin 1415, U.S. Geological Survey, 16 pp.

Romanowitz, C. M., et. al., 1970, "Gold Placer Mining; Placer Evaluation and Dredge Selection," Information Circular 8462, U.S. Bureau of Mines, 56 pp.

Sargeant, E. W., 1934, *Centrifugal Pumps and Suction Dredgers*, 3rd ed., C. Griffin & Company, Ltd., London, 202 pp.

Showers, F. S., 1969, "Gold Dredges of California; a Scrapbook, 1911-1969," Folsom, CA.

Silva, M. A., 1986, "Placer Gold Recovery Methods," Special Publication No. 87, California Department of Conservation, Division of Mines and Geology, 32 pp.

Sourdough, A., 1932, *A.B.C. of Practical Placer Mining*, Great Western Publishing Co., Denver, 62 pp.

Spence, C. C., 1989, *The Conrey Placer Mining Company, a Pioneer Gold-dredging Enterprise in Montana, 1897-1922*, Montana Historical Society Press, Helena, MT, 161 pp.

Staley, W. W., "Elementary Methods of Placer Mines," 13th ed., Pamphlet No. 35, Idaho Bureau of Mines and Geology, Moscow, ID, 28 pp.

Thompson, J., and Dutra, E. A., 1983, *The Tule Breakers, the Story of the California Dredge,* Stockton Corral of Westerners, University of the Pacific, , Stocton, CA, 368 pp.

Tourtelot, H. A., 1968, "Hydraulic Equivalence of Grains of Quartz and Heavier Minerals, and Implications for the Study of Placers," Professional Paper 594-F, U.S. Geological Survey, 13 pp.

Wells, J. H., 1989, "Placer Examination, Principles and Practice," Technical Bulletin 4, U.S. Bureau of Land Management, Washington, DC, 209 pp.

West, R. C., "Colonial Placer Mining in Colombia," Social Science Series No. 2, Louisiana State University Press, Baton Rouge, LO, 159 pp.

Wilson, E. B., 1903, *Hydraulic and Placer Mining,* J. Wiley & Sons, New York, 234 pp.

Index